ENCODE

Praying with the Language of Glory

N. DESIREE GARLAND

Amazon Kindle Direct Publishing

First Paperback Edition 2020
First Digital Edition 2020
Amazon Kindle Direct Publishing

Interior Design by Robert Smyth
Cover Design by Ronald Cash and Carol Phillips

For information regarding special discounts for bulk purchases, please contact Desiree at desiree.garland78@gmail.com. For more information about the author, visit desireegarlandministries.com.

———————————————————————————

Library of Congress Cataloguing in Publication Data

Garland, Nasal Desiree
  Encode: Praying with the Language of Glory
  1. Nonfiction—Religion—Religion & Science
  2. Prayer
I. ENCODE
LC 2020921174

———————————————————————————

PAPERBACK ISBN: 978-1-7359998-0-7
E-BOOK ISBN: 978-1-7359998-1-4

We live, fully identified
in the
NEW CREATION
Renewed in knowledge
According to the
PATTERN
of the EXACT
IMAGE of our
CREATOR
(Colossians 3:10)

CONTENTS

INTRODUCTION

Throughout my life, I have been a very spiritual person with deep faith and profound love for God. I am also very interested in science and passionate about how things work. Ever since I was a little girl, I have needed to understand how and why things happen through the lens of science.

I have always thought of science and faith as being connected, but I have learned that many people do not see this connection. I have found that those who are faith-minded sometimes ignore science, and those who are science-minded sometimes ignore faith. One of my biggest desires has always been to show others how God designed science and faith to work together, so I decided to write *Encode: Praying with the Language of Glory*.

My journey into writing this book began with one word: *metamorphosis*. During one of my quiet times with the Lord, He gave this word to me. There was nothing unusual about this situation because God often speaks just one word or phrase to me during prayer, and I may not hear more from Him about that word or phrase for quite a while. But I do know that when God speaks, I must listen and follow.

I researched the word *metamorphosis* and learned it comes from the Greek word for transformation or transforming: *metamorphoō*. This is the term used in the New Testament to describe the Transfiguration of Christ.

That discovery was intriguing to me. I now understood that God wanted me to read about the transfiguration of Christ, so I turned to some passages in the New King James Version of the Bible that described this event.

> And it came to pass about eight days after these sayings, that He took Peter, John and James and went on the mountain to pray. As he prayed, the fashion of his countenance was altered, and his raiment was white and glistering.
> —Luke 9:28-29

> Now after six days Jesus took Peter, James, and John his
> brother, led them up on a high mountain by themselves; and
> He was transfigured before them. His face shone like the
> sun, and His clothes became as white as the light.
> — Matthew 17:1-2

These passages explain that through prayer, Jesus transfigured before his disciples' eyes. During prayer, his body emitted light that was brighter than the sun.

Of course, being me, I had to figure out the science behind what had happened to Jesus. For a reason I didn't entirely understand at the time, I had to know what part prayer played in the transfiguration of Christ in the flesh and the subsequent revelation of His glory on earth.

I wondered how much energy Jesus's body had to emit to shine "like the sun," and where all of this energy was coming from. I needed to understand the mystery of what genetic language had been encoded into Jesus's DNA that activated 60 trillion cells and 7 octillion (7,000,000,000,000, 000,000,000,000,000) atoms to undergo an intense particle transition to emit light that was brighter than the sun.

The sun emits as much energy as precisely 2.88 quadrillion light bulbs. That's like giving every single person on the planet (7 billion people) a light bulb that will shine bright for their entire lifetime. How could something like this happen in a physical body?

I kept hearing the term *Bioglory* in my spirit. I did not have a clue what it meant, but it seemed to be some sort of manifestation of His glory, so I thought I should explore it a bit more and gain a better understanding.

What I learned is that *Bioglory* is not an official term or doctrine used by religious or ecclesiastical authorities today, but the concept is essential to our sacredness as human beings. We are made in the image and likeness of God. He has created our bodies to harness *Life* through *Light*.

What does that mean? The message we have from Jesus is that "God is Spirit and He is Light." The Bible doesn't say that God *has a* spirit or that He *has a* light. Jesus stated that God *is* Spirit and He *is* Light. That Light is Bioglory, and understanding the Bioglory of God is vital to connecting with the resource from which God meets your needs and moves in your life: "But my God shall supply all your need according to his riches in glory by Christ Jesus" (Philippians 4:19).

Jesus prayed that we would be glorified in the same way that the Father had glorified Him. The purpose of this book is to show you how we can all open ourselves to the true Bioglory, the Life-Light of God. "But as many as received him, to them gave the power to become the sons of God, even to them that believe on his name" (John 1:12).

So, how do we open ourselves to God's Bioglory? Many of us don't realize that when we pray, we are spending countless hours in the presence of a God who emits the highest form of Life-Light information. This information affects us in profound ways. When we pray, the same genetic language that created everything on earth is transforming us at a molecular level. It's no wonder we become more like Christ when we pray. We become transfigured through prayer. We literally start to "become like Christ."

How do I know this truth? I am not an expert on prayer, but I am a student of the Holy Spirit. In fact, I feel like such an infant when it comes to prayer, and I praise and thank the Lord of Glory for His patience with my limited understanding and application of this divine discipline. That said, I've studied this topic extensively, and I want to share my newfound knowledge with you.

I grew up hearing the word of God. When I was eight years old, I had an amazing experience while I was in church with my family. I felt the presence of God on the inside of me while I prayed. This experience was both mind-boggling and life-changing, and afterward, I began to feel and understand God differently. I no longer viewed Him as just a God who headed our church. Instead, I viewed him as the Creator of the universe.

The more deeply I thought about creation, the more I realized that the whole universe had been hardwired by God, and that His Glory affected every atom, every element, and every molecule. When I learned about DNA in middle school, I realized that everything has a digital code or genetic code that functions because of Him, including my own DNA. I understood that my DNA was programmed, and that God programmed it. Then I came across a scripture that tied all of these discoveries together for me: "Before you were born, I knew you" (Jeremiah 1:5). I was amazed. Because of my DNA programming, God knew who I would become before I was physically formed in my mother's womb.

Years later, I further connected the dots and I started to understand that Bioglory is God's genetic language that has been encoded within us. Our DNA has been programmed to accept God's Life-Light through the power of prayer.

I often hear people talking about how prayer transforms them. This feeling of transformation is more than just a feeling; transformation actually occurs on a molecular level. In this book, I will use scientific studies to back this claim and to explain how this transformation occurs.

When you pray, your DNA is being reshaped. When you take in God's Bioglory and Life-Light through prayer, you allow God to reprogram your DNA so you can become a better person. When you bring God's Bioglory into your soul, you can't hate anymore. You treat your spouse differently. You love your kids differently. Because when you pray and absorb His Bioglory, something essential changes within you.

When you pray, you become more like Christ Jesus. You stop condemning yourself because condemnation of Jesus is impossible. When you hold onto this new knowledge, you are going to let go of the old you, just like a caterpillar lets go of its old self to become a butterfly. Prayer is your cocoon state that transforms you into God's beautiful butterfly.

When you pray, God reprograms you. He literally encodes your DNA to become more like Christ. The word *encode* means to convert information into a particular form. In this case, the information being converted is God's Bioglory, which reprograms our DNA to make us better.

Our DNA passes God's Life-Light information to our body at a cellular level, so when we encode God's Life-Light, we promote healing from the core of our being. Because our spirit, soul, and body are interdependent and integrally connected, how we spend time in God's presence affects how we feel physically and emotionally. Therefore, if we believe in the power of healing as we pray in His presence, our body receives His information and responds accordingly. We have the power to accept His Life-Light and allow healing to occur. In fact, this is what prayer-encoding Bioglory is all about: healing through prayer.

God has promised to fill the earth with the knowledge of His glory. As believers all around the world continue to make themselves available, the Holy Spirit will pour out more of God's Bioglory. I hope this book will inspire you to begin your own study into the unfathomable depths of prayer-encoding Bioglory. So much is hidden in God, and if you seek Him with your whole heart, He will reveal Himself to you.

Those believers who are spending time with God in prayer are witnessing a major upsurge in the manifestation of healing and miracles. Many are longing for a restoration of the miraculous ministry of Christ, but Christ is longing for a people who will humble themselves and pray and seek His face. Miracles, signs, and wonders will automatically follow those who pray.

I pray that those who read this book will catch the same spirit of excitement I feel about the idea of being glorified by the Father through prayer. I'm convinced that until we get serious about the purpose of prayer, we won't know the half of what the Lord desires for our lives.

I wrote this book to help you reap the benefits of praying and spending time with God. I know that if you understand that you are being transfigured literally at a molecular level, you are going to keep returning to that place to commune with God. You are going to want to meditate on Him. You are going to want to transform. And you are going to make time in your busy schedule for Him.

My hope is that this book will inspire you to embrace the practice of prayer-encoding Bioglory: seeking the Lord in meaningful prayer, capturing all that God is, and encoding His Word into your life. Let these pages challenge you to more deeply understand the Word of God and to see how the connection between science and prayer can stretch and transform your mind to better appreciate God's plan—from the beginning of time to the end of time.

> For then I will restore to the peoples a pure language,   That they all may call on the name of the Lord, To serve Him with one accord.
> —Zephaniah 3:9

CHAPTER 1
DNA AS A SPIRITUAL CONDUIT

Before I formed you in the womb, I knew you.
—Jeremiah 1:15

You've probably heard of a molecule called DNA, otherwise known as "the blueprint of life." DNA (deoxyribonucleic acid) is the hereditary material in all living things. Each cell in your body contains the same DNA within its nucleus. Cells differentiate into various cell types when different genes within the DNA are switched on.

Are genes our identity? It might seem like that is true, but in fact, it is only partly true. It has been discovered that only 2 percent of your DNA genes are coded for your identity. The other 98 percent of your DNA is known as "junk DNA." Scientists often describe the junk DNA as being filled with "potential." The junk DNA carries the information that regulates what is being coded. In other words, 2 percent is what you are; 98 percent is what you are to become.

The Bible says, "And those that receive Him, he gave them the power to become the sons of God" (John 1:12). The 98 percent of non-coded DNA is to help you transform. As you spend time in God's presence, His Life-Light provides the information to be encoded within this malleable 98 percent.

This concept of junk DNA first came to light through the work of the Human Genome Project (HGP), an international team of molecular biologists, cryptanalysts (people who break secret codes), linguists (people who study languages), and physicists who worked from October 1990 to April 2003 to sequence and map the human genome. Their work gave us the ability to read the complete genetic blueprint of a human being, which brought to light this concept of junk DNA.

In studying the human genome as a group, they found they do not yet know how many genes are in the human body, and while they are making progress, they cannot be sure that they have discovered all human genes and transcripts (information codes) that make up our identities.

Additionally, they discovered a stunning inventory of regulation switches or gene enhancers embedded throughout the entire length of the 98 percent of junk DNA. The human genome has more genes that are "switched off" than any other known organism. One of the leading researchers, Ewan Birney, proclaimed, "People always knew there was more there than protein-coding genes. It was always clear that there was regulation (switches or enhancers). What we didn't know was just quite how extensive this was."

In this chapter, we will look more closely at the potential of junk DNA and its purpose in God's plan for you.

Genes Don't Act by Themselves

Genes are essentially just blueprints. In order to use these blueprints, you have to switch them on. Researchers all over the world are confirming the extensive capabilities of these switches to turn on and turn off genes within the human genome.

The field of epigenetics studies how the development and functioning of spirit, soul, and body are influenced by forces operating outside the DNA sequence, including intracellular, environmental, and energetic influences. These forces are called *epithetic signals*.

Now let's look at all of this from a spiritual perspective. DNA is the same within all living organisms. What makes the difference between a spiritually enlightened individual and a spiritually confounded individual? The information encoded within the 98 percent of junk DNA. The encoded information regulates and controls one's physical, mental, and spiritual disposition.

DNA Scroll

Let's begin at the point at which life begins in the womb, when a man and a woman come together and sperm joins with egg. Science has made it possible to capture the initial encoding and assembling of information into DNA under an electron microscope. As the sperm cell makes contact with an egg, there is a flash of light. Scientists call this moment the entrance of energy. I want to reframe it differently and call it the *DNA spiritual conduit*. This is an opening to the realm of eternity, to make way for your spirit to become encoded in the womb.

Light carries information, and I believe that this flash of light comes loaded with your unique information. I call this information your *DNA scroll*. The flash of light that occurs when sperm meets eggs combines your physical genetics with your spiritual genetics to construct you into a living, breathing being. Once the information within your DNA scroll is unveiled, the weave of your unique fabric is revealed.

Remember the passage I shared at the outset of this chapter? It helped me discover that God plays an active role in coding who we are, even before we are born:

> God told Jeremiah, "Before I formed you in the womb I knew you; Before you were born I sanctified you; I ordained you a prophet to the nations."
> —Jeremiah 1:15

God alone knows our specific genetic information, and He alone has ordained our genes to express His Bioglory.

God knew you prior to your physical manifestation. He has encoded information into your spirit at the subatomic level. When your spirit comes alive in Christ, the Holy Spirit uses this information to develop you into a child of God. This genetic information is not random. The billions of atoms that constitute the original genetic codes within your DNA come together to construct a highly sophisticated biological library of coherent, intelligent data. Your spirit can hold countless gigabytes of genetic information, which your cells can replicate. It is proven that the DNA double helix strands of just one cell are comprised of approximately 150 billion atoms, all held together through the mystery of the encoded information unique to your atomic bonding.

In Psalm 139:13-16, David shines light on how his unique atomic bonding came together by God's ordained genetic information:

For You formed my inward parts; You covered me in my mother's womb. I will praise You, for I am fearfully and wonderfully made; Marvelous are Your works, And that my soul knows very well. My frame was not hidden from You, When I was made in secret, And skillfully wrought in the lowest parts of the earth. Your eyes saw my substance, being yet unformed. And in Your book they all were written, The days fashioned for me, When as yet there were none of them.

This passage further elaborates on what God told Jeremiah. God is with us from the very beginning, bestowing upon us and coding into us the material that makes us who we are.

Whenever we study prayer-encoding Bioglory, we must understand that God has placed pre-programmed information within us, and that information can only be accessed by the Holy Spirit. At the subatomic level, the atoms that carry your information, divine abilities, and traits are either being manifested within the material world *or* they are inactive. Your DNA is either manifesting all that God has predestined for you *or* you have them shut off and are therefore not maximizing your potential.

You have more potential than you think. Two percent of your DNA has materialized, but 98 percent of your DNA is responsible for encoding the genetic language to switch on the gene for you to become, as described in Roman 8:18-19, the sons and daughters of God! You will never know the full potential of the other 98 percent of your DNA unless you spend time with God in prayer.

For I know the thoughts that I think toward you, saith the LORD, thoughts of peace, and not of evil, to give you an expected end.
—Jeremiah 29:11

God has an expected end for you, but you have to allow the Holy Spirit to activate the atoms that carry your unique genetic information—the information that was pre-ordained within you. All that you have received from God will be manifested into your reality when you turn on your atoms though prayer.

When you are in a spiritually deactivated state, things that you cannot see will continue to exist in the invisible world, but they will be out of reach until your prayers call them forth. When these atoms are activated by your faith, the information stored in your DNA will become the building blocks of your reality and potential.

I believe any number of scenarios are possible when gene expression is switched on, depending on what information your faith has activated in prayer at the subatomic level. Remember that 98 percent of your DNA is nothing more than spiritual conduits waiting to encode information that will bring your divine nature into its full potential. Think of DNA as a portal, gate, or quantum pointer to the spiritual world. You have more genes that carry information than you have genes that are currently coding your information. "Lift up your heads, O you gates! And be lifted up, you everlasting doors! And the King of glory shall come in. Who *is* this King of glory? The LORD strong and mighty, The LORD mighty in battle. The Lord of Host, He is the King of Glory" (Psalm 24:7-10). DNA is the ancient gate through which the King of Glory can enter.

The DNA Channel to God

Your DNA was designed to be a free-flowing channel for God's unlimited power and glory. Instead of thinking in a linear way that there is a compartment or a box where your DNA is closed up within your cells, think of your DNA as a spiritual conduit.

A *conduit* is simply the channel or means through which something moves from one point to another or is converted from one form to another. Conduits can be airwaves for radio or television communication, pipes for water, or wires for electricity. According to the Bible, the human body is the greatest conduit for God's Bioglory. Your DNA is a long-term conduit for His genetic information.

Have you ever flipped a switch and wondered how it makes a light turn on? The switch is a conduit that sends a message through an electrical impulse that we perceive to be electrical energy, and thus, the light is turned on. This same principle applies to our DNA as we undergo our transformation. The message from God's Bioglory is sent to the 98 percent of our non-coding DNA. His Life-Light turns on genes. God wants us to maximize our potential to fulfill the purpose He has for us. Just like the light will not illuminate the room when the switch is turned off, the 98 percent of our DNA cannot function at our optimum level without being turned on by God.

The purpose of man is nowhere more clearly revealed than in the extensive potential of the junk DNA inside each cell. What will we look like when the genetic language of Bioglory Life-Light has completed its encoding work within us? God said, "Let light shine out of darkness." He made His light shine in our hearts to give us the knowledge of His glory. In other words, we will become Christ-like. Ninety-eight percent of DNA is responsible for encoding the genetic language we need in order to become children of God!

> We stand fully identified in the new creation renewed in knowledge according to the pattern of the exact image of our Creator.
> —Colossians 3:10

Without understanding what we know about DNA today, the writer of Colossians was able to grasp and share that God created us using "the pattern of the exact image of our Creator"—essentially, by encoding us with His own DNA.

I wholeheartedly concur that it possible to engage so deeply with God that we not only experience a metamorphosis of our situation, but we also experience a metamorphosis of our DNA to the degree that our cells encode and reproduce the blueprint of Christ as we take on His appearance from the inside out. The glory that Christ has encoded within you is greater than your yesterday. It is also greater than your present circumstances and situations. This information has modified your DNA with an eternal purpose (Romans 8:18, Matthew 16:18).

Information from our external and internal environments can modify our genes and create inheritable patterns of chemical markers on our DNA.

This means that the information that has modified your genes during your lifetime can be passed down to future generations, just as information that modified the DNA of your parents and grandparents has been passed down to you. The information within our DNA can impact the next three to four generations. Consider the following scriptures:

> The Lord . . . visits the iniquity of the fathers on the children and the children's children, to the third and the fourth generation.
> —Exodus 34:6-7

> These commandments that I give you today are to be on your hearts. Impress them on your children.
> —Deuteronomy 6:6-9

> Behold, I was shapen in iniquity; and in sin did my mother conceive me.
> —Psalms 51:5

> And I will establish my covenant between me and thee and thy seed after thee in their generations for an everlasting covenant, to be a God unto thee, and to thy seed after thee.
> —Genesis 17:7

> For the promise is unto you, and to your children, and to all that are afar off, *even* as many as the Lord our God shall call.
> —Acts 2:39

> But as many as received him, to them gave the power to become the sons of God, *even* to them that believe on his name: Which were born, not of blood, nor of the will of the flesh, nor of the will of man, but of God.
> — John 1:12-13

These scriptures show that DNA modifications are passed down from generation to generation. It is scientifically proven that how we think and act is passed on to the next generation. To that end, our prayers also pass on to the next generation because our prayers are constructed of our thoughts, words, feelings, and emotions. I think you can pass on blessings through prayers and you can also pass on curses by not being prayerful or abdicating yourself from the presence of God.

Our DNA changes moment by moment, by how we direct our thinking and our words. The choices we make, the prayers we pray—all of this encodes our DNA, which we then pass on to our children and our children's children.

> Now when He was asked by the Pharisees when the kingdom of God would come, He answered them and said, "The kingdom of God does not come with observation; nor will they say, 'See here!' or 'See there!' For indeed, the kingdom of God is within you."
> —Luke 17:20-21

How is the kingdom of God within us? The process of prayer-encoding converts Bioglory into genetic instructions of Life-Light, so it is possible for the receiving cells to store and duplicate the exact blueprint image of Christ in the human genome. Prayer-encoding Bioglory, then, is Jesus's vision of the kingdom of God within you.

The Creator is working to bring us back to the original DNA blueprint of our design and to restore the kingdom of God within us. We work with Him by allowing Him access to our DNA through prayer.

CHAPTER 2
DNA'S CAPACITY TO STORE BIOGLORY

For it is you who light my lamp; the LORD my God lightens my darkness.

—Psalm 18:28

The idea that one could encode Bioglory into biological DNA might be hard to imagine, but we see that it is entirely possible according to the Bible and science. With science, this happens, as we previously covered, through epigenetics. Essentially, we have switches that dictate what gets encoded into our DNA. The Bible also explains this phenomenon.

Moses spent forty days in constant communion with God. During that time, his cells absorbed and stored so much of the genetic blueprint of God's Bioglory that they began to duplicate His exact image. When Moses came down from Mount Sinai, he was not aware that his face was radiant because he had spoken with the Lord. When Aaron and all the Israelites saw Moses, they were afraid to come near him (Exodus 34:29). His DNA was responding to the intense energy field that surrounds God, Who emits Light (Revelations 4:3, Ezekiel 1:28).

Through prayer, Moses became illuminated, literally, by the effulgence of God. The sum total of God was made known unto Moses. Not doctrinally but experientially. Doctrine without experience and transfiguration is dead religion. The title of this book, *Encode: Praying with the Language of Glory*, is about readers taking action to encode God's Life-Light through the practice of prayer. As His followers in the latter days, we are called to experience a transformation at an even greater level than Moses did on Mount Sinai (2 Corinthians 3:1-18, Haggai 2:6).

The prophet Ezekiel described God appearing as lightning in Ezekiel 1. We know that lightning is a flow of electrons creating electricity. It is extremely bright and hot—actually hotter than the surface of the sun.

When God enters a space, all of the atoms in that atmosphere become fully energized by God's Bioglory:

> When they lifted up their voice with the trumpets and cymbals and instruments of music, and praised the LORD, *saying: "For He is* good, For His mercy *endures* forever," that the house, the house of the LORD, was filled with a cloud, so that the priests could not continue ministering because of the cloud; for the glory of the LORD filled the house of God.
> —Chronicles 5:14

As the above passage indicates, when He enters a space, His Bioglory bombards the atoms and everything is filled with His Life-Light. In this chapter, we'll explore how prayer embeds His Light within us. By allowing Him to reprogram our DNA, we can store and duplicate that Light through the Spirit of Christ.

DNA and the Spiritual Realm
Russian biophysicist and molecular biologist Pjotr Garjajev placed DNA inside a quartz container and zapped it with a laser. The DNA acted like a light sponge, somehow absorbing all the photons of the light and storing them for thirty days inside a corkscrew-shaped spiral. In other words, the DNA pattern altered to store the exact blueprint of the laser light.

Gariajev's work suggests that some unknown force is holding the light in place.

One possible explanation is that the DNA is responding to an external energy field. This energy field is exchanging information with the cells in the form of light. In essence, DNA is absorbing and storing light from the field by some unknown process. This research even suggests that all living biological systems have the exact blueprint of our physical bodies stored in a spiritual field of light. Therefore, it is the spiritual energy field that is organizing, shaping, and forming the DNA blueprint pattern.

This is arguably one of the most significant scientific discoveries in modern history. It shows us that DNA has a relationship with the spiritual realm, a discovery that mainstream scientists have not yet accepted or studied in depth.

We don't often think about light as something that can be stored and duplicated as a genetic language. If Dr. Garjajev was able to capture and store light within DNA, with a mysterious quantum field that held the light in its exact shape for thirty days after the molecule itself had been removed, then it is not farfetched for us to understand that DNA is the catalyst by which the human spirit is renewed, energized, and restored by the Bioglory of God.

DNA has a limitless capacity to receive, store, and duplicate the Bioglory of God within the human genome. John caught a glimpse of this technology when he wrote, "And the Word became flesh and dwelt among us, and we beheld His glory, the glory as of the only begotten of the Father, full of grace and truth" (John 1:14).

DNA encoded the genetic language of the Word into human flesh, and Christ was born. DNA encoded the resurrection power of God within Christ. And DNA encoded the genetic language to dematerialize and rematerialize Jesus's body as he passed through a wall after the resurrection (John 20:19-31). In addition, on Pentecost, DNA encoded the genetic language of the Holy Spirit, known as the "tongues of fire," within the apostles (Acts 2).

You are not only comprised of matter. Man was created from dust. Genesis 2:7 says, "And the Lord God formed man from the dust of the ground, and breathed into his nostrils the breath of life; and man became a living soul." God breathed His energetic frequencies into man. The energetic frequencies of God are what separates man from all other created beings. We are made up of subatomic particles, elements, and structures that are vibrating at a certain frequency inside us. Quantum mechanics has confirmed that the elements and structures that we are made up of "blink" or vibrate in and out of this physical reality. Remember, DNA is a spiritual conduit, which means our bodies are capable of receiving His Bioglory.

Think about what happened to Enoch. His DNA received and stored so much of God's Bioglory that he was "translated" into the spiritual realm and did not see death (Genesis 5:24). When we pray and believe what

God has revealed to us, we are harmonizing with His energetic frequencies. When frequencies are harmonized with each other, their vibrations increase. Though we most likely won't reach the point where we are translated out of this physical reality, we can enjoy the physical and spiritual health benefits of these harmonized frequencies.

Scientists have discovered that light is energy that vibrates, and those frequencies are capable of unlocking energy fields in and around the human body, rebalancing the spirit, soul, and body and realigning the surrounding energy fields. With that in mind, I theorize that every mental illness and disease is rooted in a DNA deficiency. DNA requires God's Bioglory for vital immunity, normal physiology, and metabolism.

Scientific research has proven that all living creatures have a bio-electric energy field, an aura or light-energy, in and around them. Is it possible that this field has been diminished by sin and will continue to deteriorate until the soul dies?

Many different cultures embrace this idea, so it's not a new revelation or theory. What is new is the understanding that the primary role of the 98 percent junk DNA within us is electromagnetic reception and transmission. We can alter our body's energy fields by communicating with God through prayer and accepting his Bioglory.

Bioglory Garment

The human spirit, soul, and body need to be recharged continuously with the Life-Light of God, which comes to us through prayer. Modern medicine makes use of electromagnetic technologies, but this knowledge seems rarely explained to the people who need the information most.

Let's talk about that aura, or, for the purposes of this book, your Bioglory garment. Your garment is directly linked with your health, mental disposition, and your ability to generate and transmit light waves (bio-photons) out into your environment. As you pray and spend time with God, the Life-Light frequency in your garment strengthens.

> Arise, shine; For your light has come! And the glory of the LORD is risen upon you. For behold, the darkness shall cover the earth, And deep darkness the people; But the

LORD will arise over you, And His glory will be seen upon you.

—Isaiah 60:2

As Isaiah reinforces, God's glory is seen upon you in your garment; in the strength of your aura.

If we can only realize that Bioglory is essential to everything, including what is happening in our bodies, then Christ, the Life-Light of the world, can heal every disease and bring true homeostasis.

Evil auras are made up of low, distorted frequencies of energy. Christ has all the energy of Life and extremely high frequencies of Light. Satan, along with one third of the angels who chose to follow him, have distorted frequencies. When evil energy interacts with your Bioglory garment of Light-Life, the evil is dismantled by the energy emitted from your garment or aura. Philippians 2:15 says, "That you may become blameless and harmless, children of God without fault in the midst of a crooked and perverse generation, among whom you shine as lights in the world." We have the greater power within us. Your aura or Bioglory garment is a greater force than any energy in this world because God is within you. And God is much more powerful than Satan, who is of this world.

First John 4:4 says, "You are of God, little children, and have overcome them: because greater is he that is in you, than he that is in the world." You have the Greater One living on the inside of you. As you go throughout your day, you must remain harmonized with His Life-Light. The energetic frequencies of the Holy Spirit will multiply and manifest the nature of Christ within you. Whatever energy is vibrating on the inside of you will be felt and seen by others in the atmosphere, because energy can create an image before your words or actions do. The energies you continue to harmonize with will become a part of your presence. Energy creates presence that is hardwired into your persona.

Your presence can consist of Life-Light energy or dark, distorted energy. Satan will do all that he can to keep you away from the presence of God, because he knows that God's Bioglory can transfigure your persona from the inside out. Have you ever noticed when the atmosphere shifts for the positive when someone walks into a room, even though that person has not spoken a word? It is because there is a greater Light-Life resonating from that person than the negative energies present in the room. Second Corinthians 4:16 tells us this: "Therefore we do not lose heart. Even though our outward man is perishing, yet the inward *man* is being renewed day by day." In addition, John 1:5 reads, "The light shines in the darkness, and the darkness has not overpowered it."

DNA Math is Quantum

Our bodies operate off of stored energies, and luckily, DNA math is quantum. According to the Merriam-Webster Dictionary, the word *quantum* means "the very small increments or parcels into which many forms of energy are subdivided." Quantum mechanics studies how these units of energy work together. Our bodies have the capability to continuously multiply energy. I have always said that anyone who spends time with God can shift any atmosphere because DNA has the capacity to receive, store, and duplicate Bioglory according to the theory of quantum mechanics. When our DNA is filled with Bioglory, the radiant energy of the Holy Spirit flows from us. The devil's kingdom is no match for a person who spends time with God in prayer. The key is knowing that God's Bioglory is in your DNA unlocking or "switching on" gene expressions. The genetic language of Life-Light is the kingdom of God within you that has been encoded by the Holy Spirit when you received Christ.

The Word says that one can put 1,000 to flight, and two can put 10,000 to flight. Run that calculation a little further and you will see that three can put 100,000 to flight; four can put 1,000,000 to flight; five can put 10,000,000 to flight; and so on.

What does "putting to flight" mean in this context? It means that when you show up filled with the Bioglory of God, Satan and his fallen angels have to flee. They cannot co-exist in an atmosphere that is filled with the substance of God. Why? Because Satan and his fallen angels, powers, and principalities are filled with negative vibrations of distorted energies and dysfunctional evil natures. We who are children of God are children of the Light. And darkness cannot exist in the presence of light.

Jesus said, "I am the light of the world; he who follows Me will not walk in the darkness, but will have the light of life" (John 8:12). Insight, mindfulness, and wisdom come through meditating on God's truth. Let's take God's Word and meditate on His truth.

> This Book of the Law shall not depart from your mouth, but you shall meditate in it day and night, that you may observe to do according to all that is written in it. For then you will make your way prosperous, and then you will have good success.
> —Joshua 1:8

As Joshua explains, by meditating in the Book of Law, by praying to God and allowing His Life-Light to flow through you, you will achieve prosperity and success.

I want to shape my world with the knowledge of God's glory, with the truth of His Word, because I know, as a human being and a believer, I could perish from a lack of knowledge. We all need knowledge of God's glory in order to transform. "For it is the God who commanded light to shine out of darkness, who has shone in our hearts to *give* the light of the knowledge of the glory of God in the face of Jesus Christ" (2 Corinthians 4:6).

The more knowledge we receive from God, the greater the weight of His glory we receive, and the more our garment or aura will be reinforced. The Life-Light of Christ is in the knowledge, and the knowledge of Christ unlocks the Bioglory of God within us.

As discussed in the previous chapter, the weave of your unique fabric is revealed within your DNA scroll. God knew you before you were formed in your mother's womb, and He has locked genetic codes or knowledge inside of you—we were made in His image and likeness. There is knowledge within the 98 percent of your junk DNA that hasn't been unlocked. Second Corinthians 4:7 says, "We have this treasure in earthen vessels [our bodies], that the excellence of the power may be of God and not of us."

We learned that genes carry the information that determines your traits, and that this information can be modified. Therefore, by bringing in the correct information, you can shape your identity to change your destiny.

God shaped the identity of the universe with His Word. Moment by moment, day by day, you can modify your DNA by choosing to speak faith-filled words. The energy in these words are encoded signals that get captured by your DNA and networked through your body.

Understanding how He has designed our DNA to be superconductors of His Bioglory genetic language of Life-Light gives us a greater sense of purpose. Remember that 98 percent of our DNA is doing something far greater than coding protein. This 98 percent modifies our genetic makeup by the minute.

Our thoughts, words, and actions send signals to our DNA to switch genes on or off. Each time our DNA is modified, that modification is stored as a genetic code, which gets passed down to later generations. Science is now revealing what the Bible has taught us all along:

> Because of Adam's sin, we were all condemned to perish, no matter how many good things we did or how well we obeyed the Law in order to be godly.
> —1 Corinthians 15:22

Adam's modifications may have switched off many of your genes, but you can switch them back on by accepting Christ as your Lord and Savior through prayer. "For as in Adam all die, even so in Christ all shall be made alive" (1 Corinthians 15:22).

CHAPTER 3
BIOGLORY MORPHIC RESONANCE

𝕀𝕏𝕀𝕏𝕀

The Spirit of God, who raised Jesus from the dead, lives in
you. And just as God raised Christ Jesus from the dead, he
will give life to your mortal bodies by this same Spirit living
within you.
—Romans 8:11

Christ in the flesh has taken the human genome (DNA) beyond the limitations of space and time, the mind, and the physical body. We must live
a life that is fully identified in this knowledge as we are being transformed
according to the pattern of the exact image of our Creator.

The very same Spirit that is resonating within Christ in heaven is
resonating with us on earth (Matthew 6:10). The Holy Spirit contains
information about the form of the whole body of Christ, as well as what
genetic codes should be encoded into our cells. "For in him we live, and
move, and have our being" (Acts 17:28). This is Bioglory Morphic Resonance. In this chapter, we'll explore how this phenomenon relates to
prayer.

Understanding Morphic Resonance
According to British biologist Rupert Sheldrake, *morphic resonance* is the
concept that an individual organism can be influenced by the behavior
of another organism of the same species because there is an "inherent
memory within life"—a knowledge that exists in energy fields, or morphogenetic (form-shaping) fields, around and within organisms.

Sheldrake says that the field is characterized by inherent or natural
vibrations, known as *resonant frequencies*, that hold information about
the organism's potential abilities and forms, which he describes as being
analogous to genetic information. The information is stored in the field
and is communicated through vibrations. When the fields of two similar

organisms interact with each other, the resonant frequency of the one induces the second field into vibrational motion until the two fields resonate with each other at the same morphic frequency. This scientific discovery is consistent with the Bible. The Holy Spirit has the remarkable ability to create Bioglory Morphic Resonance Fields, arranging lines of force to shape and bring order to systems of all levels of complexity: atoms, molecules, cells, organs, and organisms. On a larger scale, it can mold ecosystems, solar systems, and even galaxies—essentially shaping and forming things into what the Father desires.

The opening three verses of the Bible describe how the Holy Spirit creates Bioglory Morphic Resonance in language that powerfully resonates with the science Dr. Sheldrake uncovered:

> In the beginning God created the heavens and the earth. The earth was without form and void; and darkness was on the face of the deep. And the Spirit of God was moved upon the face of the waters. Then God said, "Let there be light," and there was light.
> —Genesis 1:1-3

The earth was formless and empty. There was no light, no stars, and no sun to emit light; the entire embryonic universe was engulfed in complete darkness. Atoms, the building blocks of matter, had not yet taken any kind of form, nor had the mass of galaxies been formed.

The Holy Spirit hovered over the waters (with a vibration-producing resonant frequency), and as God spoke, the invisible atoms were fundamentally altered, and they eventually collapsed into matter. They were no longer formless.

Storage and Transmittal of BM-Field

Bioglory Morphic Fields (BM-Fields) of the Holy Spirit are interconnected systems. Everything in those systems relate both to internal and external realities, locally and nonlocally. In other words, BM-Fields are hardwired like any other system: everything is connected. For example, let's take a look at the human body, which is a system most of us understand. If one area of the body is damaged, cells will communicate with one another throughout the body in order to fix that damage. The body's hardwiring knows when a pattern is broken, and it will repair itself by re-coding the correct information. Similarly, when God spoke everything into existence by His Word, he created a pattern for the universe. The main hardwiring principle of the universe is that every part or area contains knowledge of the whole Word of God. This truth is emphasized in the phrase taken from The Lord's Prayer: "on Earth, as it is in Heaven."

This view is reflected by the founder of Turning Point Program and one of America's most trusted Bible teachers, Dr. David Jeremiah. In his book, *The Prayer Matrix*, he writes, "God has hardwired the universe to work through prayer. It's breathtaking to realize that the all-powerful God intends you to have such a huge part in the work of ushering in His kingdom for all eternity. Enter the matrix and discover the ultimate reality—a reality beyond your imagination." I will add that the hardwiring of the universe is made possible by the BM-Fields of the Holy Spirit. We can enter this matrix field through prayer. As Dr. Jeremiah so clearly states, "At the moment we pray, we become subject to the most powerful force in the universe."

Dr. Jeremiah also explains why prayer is so powerful and why consistent prayer requires so much discipline. For him, his battle with cancer a few years ago brought this problem into perspective. While he had always been a person of prayer, he prayed differently when ill. He expounds on this idea in his book: "When things are going smoothly in your life, you pray one way. When you get into a tight spot, you pray another way. Your pleas become more intense. You find yourself crying out to God and spending more time in his presence."

Dr. Jeremiah's words reinforced my belief in God's Bioglory. I knew then as I know now: you can encode God's Life-Light into your life, into your body, and into your DNA. Dr. Jeremiah had cancer, and he believed that spending time in the presence of God helped him through that tough time. He was healed because he spent time in the presence of God, but he explains that we should not limit ourselves; we need to find the time to be in the presence of God every day (not just when we are ill or in need).

As he says, "I believe that God wants us to pray earnestly like that all the time in His presence on both good days and bad days."

In prayer, the invisible elements of the earth will take on form and shape and materialize according to your faith-filled words. If you want to see the "desolated" or "formless" conditions of your life, ministry, business, or community changed, then prophesy the Word into these situations—to shape and bring prophetic order to all systems, at every level of complexity:

> Whatever things you ask when you pray, believe that you
> receive them, and you will have them.
> —Mark 11:24

As Mark explains, when you believe that God will answer your prayers, then God will answer them.

BM-Fields and Prayer

Every day of your life, the BM-Field of the Holy Spirit creates an atmosphere around you that is conducive to your faith-filled prayers. The invisible waves and particles of Bioglory carry the instruction of Life-Light into the atoms that form the building blocks of larger physical objects that are eventually viewed by your naked eyes. Without your intentional faith-filled prayers interacting with the BM-Field of the Holy Spirit, you will face "desolation" and "formlessness." Remember that it is your sole duty to call those invisible things into formation.

The BM-Field is a reality-altering super massive energy field. Nothing escapes the powerful BM-Fields of the Holy Spirit, which undergird the earth and the heavens. In the Old Testament, we find the phrase "the foundations of the earth" ten times. The Bible also mentions "the foundations of heaven" (2 Samuel 22:8). In the New Testament, we find the phrase "the foundation of the world" an additional ten times. So, the BM-Field is spread across the foundations of both the earth and the heavens.

When your fervent faith-filled prayers interact with the BM-Fields of the Holy Spirit, you will be able to effect a tremendous amount of change within the invisible and nonlocal realms of the universe because "the effective, fervent prayer of a righteous man avails much" (James 5:16).

CHAPTER 4
EXPANDING YOUR BIOGLORY MORPHIC FIELD (BM-FIELD)

Have you commanded the morning since your days began,
And caused the dawn to know its place, That it might take
hold of the ends of the earth, And the wicked be shaken
out of it?
—Job 38:12-13

Do you know that you can command systems and realms within your life, community, and nation to come into alignment with the Word of God? Romans 4:17 tells us that we serve a God who has the power to "quicken the dead and call those things which be not as though they were." In other words, you were created in the image and likeness of God to do the same. When you command your morning, you are standing in agreement with the Word of God and calling those things "which be not as though they were."

The greatest technology we have is prayer. God's Word is His dunamis (strength, ability, or inherent power) in the mouth of the righteous. Although you may not see it immediately, your prayers create frequencies that set in motion a higher vibrational resonance effect, which will ultimately become so intense that spiritual glass ceilings will shatter, chains of bondage will break, and evil lines of communication will collapse—all because your prayers have created a vibrational resonance in the quantum realm. This vibrational resonance acts as a dunamis expansion of your BM-Field.

All physical and spiritual elements have a resonant frequency. When you begin to pray with God's Word, His dunamis is fine-tuning everything; the whole atmosphere in your immediate space and in the surrounding environments will buzz at a certain pitch or vibrate at a certain speed. These are some of the effects of prayer resonance. It is the mechanism by which all fields (systems) that are misaligned are induced to the same vibrational motion until they resonate at the same frequency as the spoken Word of God.

According to Isaiah 55:11-13, The Lord said "So shall My word be that goes forth from My mouth; It shall not return to Me void, But it shall accomplish what I please, And it shall prosper *in the thing* for which I sent it."

In this chapter, we will explore prayer resonance and how you can harness your BM-Fields in the spiritual realm to impact and improve your experience in the physical world.

The Walls Shook

Nikola Tesla was a genius who mastered the concept of mechanical resonance. He invented AC current electricity, triumphing over Thomas Edison's proposal to power grid cities with DC electricity. With this discovery, Tesla was called "the father of modern electricity."

In 1898, while working in his laboratory in New York, Tesla was experimenting with a small electrical oscillator that created very low-frequency vibrations. As he continued to experiment, Tesla discovered that he could use his electrical machine to set "higher frequencies" in motion to cause buildings in Manhattan to shift and shake violently. In other words, his machine could induce a man-made earthquake. Tesla had created an oscillator that set in motion a mechanical resonance, both in his immediate space and in the surrounding buildings, that would have led to the collapse of these buildings, but he quickly destroyed his machine with a hammer.

Looking at the book of Joshua through the lens of what Tesla discovered, you can understand the science behind the collapse of the walls of Jericho during the first battle fought by the Israelites for the conquest of Canaan. In this story, God taught the Israelites an important strategy. He taught them how to expand their Bioglory Morphic Field to gain an advantage over their enemy.

> Now Jericho was straitly shut up because of the children of Israel: none went out, and none came in. And

the LORD said unto Joshua, See, I have given into thine hand Jericho, and the king thereof, and the mighty men of valour. And ye shall compass the city, all ye men of war, and go round about the city once. Thus shalt thou do six days. And seven priests shall bear before the ark seven trumpets of rams' horns: and the seventh day ye shall compass the city seven times, and the priests shall blow with the trumpets. And it shall come to pass, that when they make a long blast with the ram's horn, and when ye hear the sound of the trumpet, all the people shall shout with a great shout; and the wall of the city shall fall down flat, and the people shall ascend up every man straight before him. —Joshua 6:1-5

So the Israelites followed the command and marched around the city.

And it came to pass, when Joshua had spoken unto the people, that the seven priests bearing the seven trumpets of rams' horns passed on before the LORD, and blew with the trumpets: and the ark of the covenant of the LORD followed them. And the armed men went before the priests that blew with the trumpets, and the reward came after the ark, the priests going on, and blowing with the trumpets. And Joshua had commanded the people, saying, Ye shall not shout, nor make any noise with your voice, neither shall any word proceed out of your mouth, until the day I bid you shout; then shall ye shout. —Joshua 6:8-10

By the Power of the Holy Spirit, they were creating a super-massive BM-Field around the city and fine-tuning their vibration to match the "low frequency" of the people and the city of Jericho. After the BM-Field was completed by the Holy Spirit, on the seventh day, the Israelites were instructed to then fine-tune their "vibrational resonance" to the sound of Heaven and set in motion the dunamis effects of the BM-Field.

> So the people shouted when the priests blew with the trumpets: and it came to pass, when the people heard the sound of the trumpet, and the people shouted with a great shout, that the wall fell down flat, so that the people went up into the city, every man straight before him, and they took the city.
> —Joshua 6:20

Through their vibrational resonance, and by the command of God, the Israelites brought the walls down.

As proven in this Biblical passage, the vibration resonance of your prayers and praise can do anything. As a righteous follower of God, it is your duty to step up your prayer with dunamis power to fit the natural and spiritual vibrations of the earth, so evil fortresses in high places will come crashing down. My suggestion is that you fine-tune your prayers according to your situation. If the situation is pressing, you must press back with the full power of your prayers.

Joshua and the Israelites were in a heated battle, and he fine-tuned his prayer according to the situation:

> Then Joshua spoke to the LORD in the day when the LORD delivered up the Amorites before the children of Israel, and he said in the sight of Israel: "Sun, stand still over Gibeon; And Moon, in the Valley of Aijalon. So the sun stood still, And the moon stopped, Till the people had revenge upon their enemies. *Is* this not written in the Book of Jasher? So the sun stood still in the midst of heaven, and did not hasten to go *down* for about a whole day. And there has been no day like that, before it or after it, that the LORD heeded the voice of a man; for the LORD fought for Israel.
> —Joshua 10:12-14

BM-Fields are Indestructible

As we review the scriptures, we discover that the evidence of BM-Fields in the earth realms are not farfetched. The power that the Holy Spirit displays over systems of all levels of complexity through these BM-Field is mind-boggling.

In 2 Kings 13:21, Elisha's bones cause a dead body to come alive: "Once while some Israelites were burying a man, suddenly they saw a band of raiders; so they threw the man's body into Elisha's tomb. When the body touched Elisha's bones, the man came to life and stood up on his feet." The latter part of this passage hints at the power of the Holy Spirit to convert the same energy within the BM-Fields of Elisha's bones to revive, ratify, and gird up of the loins of the church. Could it be that the power within the BM-Fields of Elisha's bones is the same power that raised Christ from the dead? And is this power now conserved within the church?

Now, here is where this theory gets interesting! In physics, the term *conservation* refers to something that doesn›t change. This means that the power within the BM-Field is constant over time. It has the same value both before and after an event. Therefore, in both life and death, Elisha's DNA (bones) carried the "double portion" of Elijah's anointing, which came upon him as a physical garment. Even in death, his bones (DNA) still resonated with the very same power.

The power within BM-Fields is always conserved; it cannot be destroyed. In essence, the Holy Spirit can constantly convert this power from one form into another. We see this phenomenon in the tangible mantle of Elijah in the Old Testament, which was converted into a super-natural BM-Field in the New Testament.

The mantle of Elisha originated from Elijah, and many scholars believe this was the same mantle that rested on John the Baptist. I believe that this mantle has been passed down throughout the history of the Old Testament as a tangible physical garment that never lost its energy value; however, in the New Testament, the Holy Spirit converted the energy of the mantle from a physical garment to a spiritual BM-Field, which engulfed and saturated the body. John was the first man in the New Testament to experience the BM-Field of the Holy Spirit as an outer mantle, then he passed it on to Christ. "And John bore witness, saying, 'I saw the Spirit descending from heaven like a dove, and He remained upon Him'" (John 1:32).

Today, my heart is overjoyed to know that Jesus passed this mantle on to the church. The very same BM-Field of the Holy Spirit that engulfed the bones of Elisha was passed on to the apostles. And now the church can use double portion anointing to do greater works for God's kingdom in the earth realm.

So as you can see, it's not happenstance that the Prophet Ezekiel also had the vision of the dry bones (DNA) receiving the conserved energy of the Holy Spirit through BM-Fields. In the Book of Ezekiel 37:1-4, the Holy Spirit uses the four winds of the earth to create another super-massive BM-Field over the dry (desolated) bones. The science of epigenetics and DNA becomes even more evident when the Word of God envelops the dry bones (DNA) as the four winds of the Spirit hover over the bones to reprogram, transform, and bring them back to life.

> God asked Ezekiel, "Son of man, can these bones live?"
> Ezekiel answered, "O Sovereign LORD, you alone know."
> —Ezekiel 37:3

DNA is always doing two essential things, *listening* for a genetic language and *responding* to it. The DNA, portrayed as desolated bones, is listening for a genetic language to encode.

> Then the Lord said to Ezekiel, "Prophesy to these bones and say to them, 'Dry bones, *hear* the word of the LORD!'"
> —Ezekiel 37:4

Ezekiel is instructed to prophesy to the BM-Field that the Holy Spirit had created with the four winds and tell the field to release breath that will resonate or vibrate within the bones (DNA).

As previously mentioned, all physical and spiritual elements have a resonant frequency. In this passage, *breath* represents Ezekiel's prayer resonance. This resonance is the mechanism by which misaligned systems are induced to shift their vibrational patterns until they resonate at the same frequency as the spoken Word of God. In this case, the dry bones are listening to hear the spoken Word so that they may resonate Life-Light.

> Then said he unto me, Prophesy unto the wind, prophesy, son of man, and say to the wind, Thus saith the Lord GOD; Come from the four winds, O breath, and breathe upon these slain, that they may live.
> —Ezekiel 37:9

The bones (DNA) encoded the genetic language, the instructions from God's Life-Light. They could no longer be described as desolate, for the bones had taken on the Life-Light of God from the BM-Field.

Life Through Prayer
Prayers grounded in the BM-Field of the Spirit encode breath into desolate situations, systems, and realms. Science shows that it was entirely possible for the dry bones to reanimate, and I believe that the Lord chose Ezekiel to exercise the power of the BM-Field over the physical and quantum world. He was given the power to probe the depths and reveal the possibilities of God's Bioglory: "So that the things which are seen [visible] were not made of things which are visible" (Hebrew 11:3).

Ezekiel saw the physical condition of the bones, then he prophesied to the BM-Field to modify the genetic condition of the bones from the subatomic realm. We must do the same. Every atom, molecule, and element are listening, ready to respond to your faith-filled words. Jesus said, "For truly I say to you, if you have faith the size of a mustard grain, you will say to this mountain, Move from here to there, and it will move and nothing will be impossible for you" (Matthew 17: 20).

We have already established that the human body itself is a resonator. Now, let's have a look at what happens to the Bioglory Morphic Field when two or three resonating bodies come together with one accord to pray. The result creates a harmonic prayer resonance within the earth, and when multiple people are gathered together in His name, the power of that resonance is incredibly strong (Matthew 18:20).

Can you image the BM-Field that was created by the Holy Spirit during Pentecost, while the apostles were all praying in the Upper Room?

> Suddenly a sound like the blowing of a violent wind came from heaven and filled the whole house where they were sitting. They saw what seemed to be tongues of fire that separated and came to rest on each of them.
> —Acts 2:2-3

Let's look at another example of the ultimate power of the BM-Field. In the Book of Luke, the angel explains to Mary that she is about to become the mother of God and this process will take place through the BM-Field of the Holy Spirit.

> "You will conceive and give birth to a son, and you are to call him Jesus. He will be great and will be called the Son of the Most High. The Lord God will give him the throne of his father David, and he will reign over Jacob's descendants forever; his kingdom will never end."

> "How will this be," Mary asked the angel, "since I am a virgin?"

> The angel answered, "The Holy Spirit will come on you, and the power of the Most High will overshadow you. So the holy one to be born will be called the Son of God.
> —Luke 1:31-35

As you can see, the Word of God has the power to alter the BM-Field and accomplish marvelous things. Through faithful prayer, you have that same power.

Christ in the Flesh

Let's take a deeper look at the case of Mary. A Bioglory Morphic Field is created by the Holy Spirit as He hovers over and covers Mary to shape and form the Life-Light of Christ in her womb—the BM-Field of "Christ in the flesh." The BM-Fields created by Christ in the flesh are the highest vibrational frequencies that the human genome system has ever coded. "For in Him dwells all the fullness of the Godhead bodily; and you are complete in Him, who is the head of all principality and power" (Colossians 2:9).

Christ in the flesh is the embodiment of the Father. All the fullness of the very substance of God dwells within the biological DNA of Christ. All sixty trillion cells within the body of Christ are filled with the genetic language of Bioglory. All that is, all that was, and all that the human genome will ever be was encoded into Christ.

The church today continues to resonate in the BM-Field of Christ in the flesh, and the spirit of the anti-Christ will continue to fight against this truth. This spirit wants the world to deny that Christ has come in the flesh and created these BM-Fields within the human genome.

In a discussion with His disciples in the Book of Matthew, Jesus asks, "Who do men say that I am?" Only Simon Peter delivers the correct answer:

> "You are the Christ, the Son of the living God."
> Jesus replied, "Blessed are you, Simon son of Jonah, for this was not revealed to you by flesh and blood, but by my Father in heaven!"
> —Matthew 16:16-17

The morphogenic fields of the church resonate from Christ. Therefore, no power in heaven or earth can destroy the church. As we've previously discussed, all DNA is the same, what makes the difference is the information encoded within it. Christ in the flesh, The King of Glory, has encoded His kingdom into your DNA. You have His information within you.

You have been given the keys to the kingdom. In both the Old and New Testaments, keys symbolize power and authority. The nature of that power and authority is within Christ. The keys of the kingdom are connected to the BM-Field of the Holy Spirit, particularly in the Body of Christ. Christ in the flesh is the source of your power and authority.

When Christ in the flesh rose from the dead, He declared, "All authority has been given to Me in heaven and on earth" (Matthew 28:18).

Jesus has all authority; His body (DNA) is the ultimate key-holder of all power and authority. He says to us, "Now you are the body of Christ, and each one of you is a part of it" (Corinthians 12:27).

Every believer has the keys. Jesus explains this truth in the book of Matthew:

> I will give you the keys of the kingdom of heaven; whatever you bind on earth will be bound in heaven, and whatever you loose on earth will be loosed in heaven.
> —Matthew 16:19

These keys are encoded within you by the Holy Spirit. It is your right to be a king and priest and unlock God's power and authority on earth, so that His will is accomplished.

The All-Encompassing BM-Fields of the Holy Spirit

Today, the church is resonating from the BM-Field of the Holy Spirit. We, as born-again believers, are held together as a body of believers in a morphogenetic field that contains an original or collective DNA blueprint memory of "Christ in the flesh," inherent to all within the all-encompassing field. "For in Him we live and move and have our being" (Acts 17:28).

As members of the Christian community, when we are Christ-like, we tap into in the collective memory of the fruits of the Spirit, including grace, healing, faith, prayer, obedience, and so on. All of these collective memories are formed from Christ, who was the first of His Kind. The BM-Field of the Holy Spirit contains the record and cosmic memory of Christ.

The BM-Fields that the Holy Spirit has created through Christ in the flesh are not a type of energy, but these fields *organize* energy. The BM-Fields are made up of parts, which are, in turn, parts of a larger whole. For example, Christ is in the Father, and the Father is in Christ. Christ is in you, and you are in Christ, and we make up "the body" that is held together by the Holy Spirit. Consider the following scriptures:

> For as the body is one and has many members, but all the members of that one body, being many, are one body, so also *is* Christ. For by one Spirit we were all baptized into one body—whether Jews or Greeks, whether slaves or free—and have all been made to drink into one Spirit.
> —1 Corinthians 12:12-13

> But the Comforter, which is the Holy Ghost, whom the Father will send in my name, he shall teach you all things, and bring all things to your remembrance, whatsoever I have said unto you.
> —John 14: 26

In reviewing the above scriptures, we understand that the BM-Fields contain memory of everything that Christ experienced while on earth and everything He is now experiencing being seated on the right Hand of the Father. The BM-Fields are nonlocal and multidimensional. We can retrieve information from the BM-Fields through prayer. In prayer, we have full access to obtain all power, authority, thoughts, and behaviors that Christ displayed during His ministry on earth.

When Christ tore the temple veil, it was a clear statement that we are one and the same because we share in the same earthly experience through the BM-Field of the Holy Spirit. "When the Helper comes, whom I will send to you from the Father, that is, the Spirit of Truth, who proceeds from the Father, He will testify about Me" (John 15:26).

The Holy Spirit is a direct link from the Old Testament prophecies of Christ to the New Testament manifestation of the Word made flesh, to the church here in the present, and to the ministry of Christ moving forward.

Jesus prayed to the Father, "The Glory which you gave Me, I have given them!" (John 17:22). Here, Jesus is referring to the DNA blueprint of His body, which resonates from the BM-Field of the Holy Spirit. The line "I have given them" is in the perfect tense as it carries the idea that the action is still occurring. Christ has given His glory and is still giving that glory every day. This glory is set forth by means of Bioglory Morphic Resonance, which creates our union with one another, with the Father, and with the Son.

There is never a time that we are not tuning into, resonating, and repeating the genetic language of Life-Light from the Archetype. He is the True Vine, and we are the branches (John 15:5). We are not second to Him, we're *one with Him*—sharing the same BM-Field of the Holy Spirit. As indicated in Galatians 2:19-20, the Father has made us spiritually alive through Christ. So what does that mean for us? Consider the following scriptures:

> Therefore, you are no more strangers and foreigners, but fellow citizens with the saints and of the household of God, built upon the foundation of the apostles and prophets, with Jesus Christ himself being the chief corner stone.
> —Ephesians 2:19-21

> But now apart from the law the righteousness of God has been made known, to which the Law and the Prophets testify. This righteousness is given through faith in Jesus Christ to all who believe. There is no difference between Jew and Gentile, for all have sinned and fall short of the glory of God, and all are justified freely by his grace through the redemption that came by Christ Jesus. God presented Christ as a sacrifice of atonement, through the shedding of his blood—to be received by faith. He did this to demonstrate his righteousness, because in his forbearance he had left the sins committed beforehand unpunished—he did it

to demonstrate his righteousness at the present time, so as to be just and the one who justifies those who have faith in Jesus.
—Romans 3:21-26

As the scriptures clearly show, we are all one with God. This connection is possible through the righteousness of Christ. This oneness brings a sense of belonging, acceptance, kingship, and security. Out of this connection comes a release of His power, and the glory of God is able to transform and transfigure our DNA.

CHAPTER 5
THE BIOGLORY UNIVERSE

I have told you these things, so that in Me you may have peace. In this world you will have trouble. But take heart! I have overcome the world.

—John 16:33

Albert Einstein believed that the universe is composed of interconnected force fields. Physicists have described some of these fields as constructs of finite reality held within a greater infinity. Because of these fields, reality is both local (here and now), and nonlocal (across time and space), which means that everything is interconnected. Given this interconnectivity, the physical body cannot be isolated or separated from these fields, but rather, it is an integral part of these fields and their spheres of influence. With this truth in mind, in this chapter we will explore the Bioglory universe.

As Christ descended, His body set in motion BM-Fields within the three-tiered universe:

1. The Earth Surfaces
2. The Six Regions of the Underworld (death, hell, the grave, the abyss, regions under the sea, and the deepest depth of the underworld)
3. The Subdivided Heavens (first heaven, second heaven, and third heaven)

Because we are actually composed of fields, we need to understand that Christ in the flesh has shaped and is still reshaping the world. His work directly affects our well-being. Christ became the driving force for humanity itself. His every word, thought, and deed reached deep into the quantum realms, creating new morphogenetic fields that are pushing the whole of humanity in a new direction—the direction of greater glory.

Earth Surfaces

Christ in the flesh, full of Bioglory and sealed by the Spirit, carried out His ministry on Earth.

> For the perfecting of the saints, for the work of the ministry, for the edifying of the body of Christ: Till we all come in the unity of the faith, and of the knowledge of the Son of God, unto a perfect man, unto the measure of the stature of the fullness of Christ: That we henceforth be no more children, tossed to and fro, and carried about with every wind of doctrine, by the sleight of men, and cunning craftiness, whereby they lie in wait to deceive; But speaking the truth in love, may grow up into him in all things, which is the head, even Christ: From whom the whole body fitly joined together and compacted by that which every joint supplieth, according to the effectual working in the measure of every part, maketh increase of the body unto the edifying of itself in love.
> —Ephesians 4:12-16

During his time on earth, His Body was vibrating at the highest frequency of heaven and creating BM-Fields for His disciples on earth, to give them direction and to carry out His work. Now let's take a look at what happened to the Bioglory within Christ's Body when He died on the cross.

We previously said that Bioglory is incorruptible, indestructible, everlasting, immortal Life-Light. Like energy, Bioglory cannot be created nor destroyed; rather, it can only be transformed or transferred from one form to another.

Six Regions of the Underworld

Christ descended into the Six Regions of the Underworld as the King of Glory to proclaim His victory over sin, death, and hell, and He rose with a material body.

Bioglory encoded His body with superior DNA, giving the flesh all its vigor and energy, all its power and activity, and all its worth and value.

These codes were also designed to resonate within the human genome once Christ was resurrected. His body was the seed and first fruit of the dead—bone of our bone, flesh of our flesh.

Christ conquered the Six Regions of the Underworld in precisely the same flesh and bones that had suffered on the cross and had been laid in the grave. But how exquisite are the BM-Fields of His Body, created by His own infinite art and skill? How transcendent are His BM-Fields, which are so refined and magnified by the power of the Spirit? Because Christ conquered the Six Regions of the Underworld, we have also conquered them, and the enemy has no power over the BM-Fields that resonates within our DNA (John 16:33).

Second Corinthians 12:2 talks about the third heaven. Logic dictates that if there is a third, there must also be a first and second.

In the scriptures, the third heaven is where God Almighty sits in dominion. Because it is the throne of God from whence He rules over the universe, it is the upper part of the universe and therefore set above all. In other words, the heaven where God dwells is placed far above all the other heavens.

The second heaven is where the moon, sun, stars, and the planets dwell—the outer space. This is where Satan rules his evil empire with legions of demons and fallen angels.

The first heaven is comprised of the six layers of the atmosphere: the troposphere, the stratosphere, the mesosphere, the thermosphere, the ionosphere, and the exosphere—which are all within the earth realm.

As He ascended, His glorified body set in motion BM-Fields in the first, second, and third heavens, realigning forces to shape and bring order to those systems.

The BM-Fields of Christ Resonate a Superior Order

Christ ascended far above all heavens to organize atoms, cosmos, planetary systems, solar systems, galaxies, and dimensions—wherever the wicked principalities, powers, and rulers of darkness inhabited. His Body resonated a superior order, creating BM-Fields. Now all who are filled with the Spirit have been given power over everything within these fields to bind and loose. God not only came down from heaven to earth, to dwell in

THE BIOGLORY UNIVERSE 43

human flesh, but God has defeated Satan, death, hell, and the grave *while in human flesh*, then ascended on High as King and Lord over all.

Because Jesus completed all these tasks while in a human body, we too will resonate from His Bioglory M-Fields. Paul understood the science of morphic resonance, and he was excited to share this knowledge with the church. In Galatians 4:19, Paul proclaimed, "My little children, for whom I labor in birth again until Christ is *formed* in you." In another translation of the same verse, he says, "Oh my dear children! I feel as if I am going through labor pains for you again, and they will continue until Christ is *fully developed* in your lives."

When the life of Christ is "formed" in us through the BM-Fields of the Holy Spirit, His power, glory, and faith is activated in our lives.

Here is how Paul further explains the revelation of the BM-Fields to the church: " I am crucified with Christ: nevertheless I live; yet not I, but Christ lives in me; and the life which I now live in the flesh, I live by the faith of the Son of God, who loved me and gave Himself for me" (Galatians 2:20).

Take note that Paul was not speaking of an experience that will take place in the future. He said, "the life which I now live in the flesh [physical body]." This is what excited Paul, and I hope you share in the same bliss. Christ in the flesh has created a morphic field within the human genome, so we can encode the fullness of God's Bioglory into our biological DNA. Consider these next passages:

> And what is the exceeding greatness of his power to us-ward who believe, according to the working of his mighty power, Which he wrought in Christ, when he raised him from the dead, and set him at his own right hand in the heavenly places, Far above all principality, and power, and might, and dominion, and every name that is named, not only in this world, but also in that which is to come: And hath put all things under his feet, and gave him to be the head over all things to the church, Which is his body, the fullness of him that filleth all in all.
> —Ephesians 1:19-21

> To me, who am less than the least of all the saints, this grace was given, that I should preach among the Gentiles the unsearchable riches of Christ, and to make all see what *is* the fellowship of the mystery, which from the beginning of the ages has been hidden in God who created all things through Jesus Christ; to the intent that now the manifold wisdom of God might be made known by the church to the principalities and powers in the heavenly *places*, according to the eternal purpose which He accomplished in Christ Jesus our Lord, in whom we have boldness and access with confidence through faith in Him.
>
> — Ephesians 3:8-13

Here, we find that John is no less awe-stricken and amazed than Paul at the BM-Fields Christ created in the three-tiered universe. John understood that a person exists in a specific location, but BM-Fields do not since their components are everywhere at the same time. Therefore, the BM-Fields that Christ has created in the three-tiered universe are still releasing information within the universe, even as the seals are opened by the Lamb of God in the third heaven.

Whatever information is written in that heavenly book would theoretically penetrate the physical universe and the church at the deepest level—its atoms, cells, systems—and even extend beyond it. Consider the following passage:

> And I saw in the right hand of him that sat on the throne a book written within and on the backside, sealed with seven seals. And I saw a strong angel proclaiming with a loud voice, Who is worthy to open the book, and to loose the seals thereof? And no man in heaven, nor in earth, neither under the earth, was able to open the book, neither to look thereon. And I wept much, because no man was found worthy to open and to read the book, neither to look thereon. And one of the elders saith unto me, Weep not:

behold, the Lion of the tribe of Judah, the Root of David, hath prevailed to open the book, and to loose the seven seals thereof. And I beheld, and, lo, in the midst of the throne and of the four beasts, and in the midst of the elders, stood a Lamb as it had been slain, having seven horns and seven eyes, which are the seven Spirits of God sent forth into all the earth. And he came and took the book out of the right hand of him that sat upon the throne. And when he had taken the book, the four beasts and four and twenty elders fell down before the Lamb, having every one of them harps, and golden vials full of odours, which are the prayers of saints. And they sung a new song, saying, Thou art worthy to take the book, and to open the seals thereof: for thou wast slain, and hast redeemed us to God by thy blood out of every kindred, and tongue, and people, and nation; And hast made us unto our God kings and priests: and we shall reign on the earth.
—Revelations 5:4-5,13

Here, we see how every system and organism in the universe at every level in nature has been affected by interactions with the BM-Fields of Christ.

Both Paul and John found themselves tasked to the utmost to imagine the magnitude and extensive nature of the BM-Fields that Christ in the flesh had created, which they knew would resonate to the church and its members. The church exists in the material world, yet we can gain mastery over the quantum world through prayer.

"It is Finished"
Christ is now seated on the right hand of the Father, and we are seated in heavenly places within Christ Jesus. Jesus is now seated in the highest position in the universe because, as he said, "It is finished." By saying these words, Jesus indicates that the debt owed by man to his Creator on account of Adam's sin is finally and forever resolved. Christ in the flesh has completed our redemption, and now, through His BM-Fields, we are

enabled to have joint-seating with Him "far above" all principalities and powers of darkness.

I believe that His words, "It is finished," set off vibrational energies on the earth. These vibrational energies create resonance just like playing a drum in a room does. The drum creates energy that can cause nearby stringed instruments, like guitars, to vibrate at the same frequency. All that Christ accomplished while on earth becomes a part of the BM-Field, and the expectation is that it will be easier for any faithful Christian to walk in His victory.

We in the church are able to respond to Christ's implication that "it is finished" through the BM-Field of the Holy Spirit. This field is our cumulative field of biological information that shapes our faith from Christ, from the first generation of sons.

For example, Christ, who is the firstborn of God, reigned over sin in the flesh because He subjected His will to God's will: "Not My will, but Your will, be done." For that reason, sons of God everywhere should be able to reign over sin in the flesh. Christ did it while on earth, and through His obedience to the Father, He has created a BM-Field for us.

The more believers that learn to reign over sin in the flesh by subjecting their will to God's will, the easier it should become for all believers to learn how to reign over sin everywhere. Likewise, if a spiritual revival happens in a prayer group in New York for the first time, the easier it should become for prayer revivals to form all around the world. Because the apostles were all filled with the Holy Spirit on Pentecost, then it should be easier today for the praying congregation to be filled with the Spirit in every nation.

These effects should happen naturally for anyone who is a partaker of Christ in the flesh. In fact, faithful followers of Christ should be able to go further spiritually than all who have gone before them because Christ's BM-Field is available for the whole human race, making the act of becoming sons of God possible for everyone.

Advancing the Morphic Field

In the Book of John, Jesus says:

Believe Me that I *am* in the Father and the Father in Me, or else believe Me for the sake of the works themselves. Most assuredly, I say to you, he who believes in Me, the works that I do he will do also; and greater *works* than these he will do, because I go to My Father.
—John 14:11-12

The same BM-Field Jesus used to accomplish His greater works is also available to us through Him. Christ's Life-Light is now ingrained in us. We are joined in oneness with Him. Just as His life reveals your potential, your life reveals His glory (Colossians 3:40).

This effect can be seen in various religious, scientific, and technological achievements. Since the first generation, the Levitical priesthood has carried the Bioglory of God within the Ark of the Covenant. Jesus came in the flesh and carried the Bioglory of His Father in bodily form, advancing the morphic field through the Gospel.

Since Jan and Paul Crouch founded the Trinity Broadcasting Network (TBN) in 1973, the morphic field has advanced to hundreds of television stations in the U.S. and thousands of other cable television and satellite systems around the world in over seventy-five countries, where the programming is translated into over eleven languages. This example demonstrates the greater works of the BM-Field of Christ in the flesh, which influences all sorts of other factors, including books, video games, music, medicine, light therapy, fashion, and more.

Again, we belong in the "high places" of the earth, the places of elevation, cutting-edge vision, and innovation. We are instructed by Christ to occupy the high places and do greater works. As it says in Matthew 8:19, "Therefore, go and make disciples of all nations, baptizing them in the name of the Father, and of the Son, and of the Holy Spirit."

As a believer, I know that if I go into the presence of God and He fills me with His Life-Light, then I am exposed to everything that God is. I am exposed to the substance of God through that Light, which carries the information of Life. As a believer, I know that I can share that Light and help others to receive and walk in the knowledge of God's glory. Doing God's work expands the Bioglory universe.

CHAPTER 6
BM-FIELDS OF THE MIND

And he shall be like a tree planted by the rivers of water, that bringeth forth his fruit in his season; his leaf also shall not wither; and whatsoever he doeth shall prosper.
—Psalm 1:3

Christians normally talk about having the mind of Christ. To me, having the mind of Christ means having direct access to the tree of life. I say this because scientifically speaking, neurons are thoughts, and neurons grow in your brain like branches. If you have the mind of Christ, it means you are thinking of things that are good, things that are pure and holy; in return your neurons, or your branches, will look like the tree of life.

Neuroelectrical pathways are a conduit for the mind of Christ, so having good thoughts flow through your neurons and the pathways of your mind is vital to your spiritual health. Having the mind of Christ means literally having a healthy brain. In this chapter, we'll take a closer look at these neuroelectrical pathways and how good, pure thinking affects the BM-Fields of the Mind.

Branching of the Brain
In her book *Think Well, Live Well Now*, Benay Behnke beautifully and simply explains that each neuron has an axon (stem) and dendrites (branches at the top). Axons and dendrites are electrical devices that carry signals of small electricity through the fields of the brain.

Similarly, in *Switch On Your Brain*, Dr. Caroline Leaf explains that the wiring of the brain looks and performs like trees, with branches and fruits. In science, these trees are called "the magic trees of the mind" because the nerve cells in the brain look exactly like trees and they produce fruit (ideas and innovations). In other words, your brain is made up of multiple fruit-producing neural pathways.

When a person is born again, Christ begins to express Himself through that individual's renewed spirit. The supernatural influence of Christ is expressed through the BM-Fields' invisible dimensions within the human DNA, and these supernatural influences resonate throughout the human mind: "He who is joined to the Lord is one spirit with Him" (1 Corinthians 6:17). Through the BM-Fields of the Holy Spirit, Christ becomes supernaturally joined to the spirit of the human being, and His mind resonates through the neurons. Each thought that flows from the mind of Christ grows as a new, healthy neuron branch within the human brain. We are literally transformed by every new neuron branch. This scientific power is called neuroplasticity. And in terms of spirituality, it is how the mind is renewed (Romans 12:2).

David proclaimed, "And he shall be like a tree planted by the rivers of water, that bringeth forth his fruit in his season; his leaf also shall not wither; and whatsoever he doeth shall prosper" (Psalm 1:3). And Jeremiah said, "For he shall be like a tree planted by the waters, Which spreads out its roots by the river, And will not fear when heat comes; But its leaf will be green, And will not be anxious in the year of drought, Nor will cease from yielding fruit" (Jeremiah 17:8).

To appreciate what it means to be "like a tree planted by the rivers of living water," as the above scriptures describe, consider this idea from the standpoint of science. When thoughts or information are being processed in the mind, it creates an electrical flow that moves through the branches of the neurons. A typical neuron has 128 branches. There are almost 100 billion neurons in a human brain, a number hard to imagine. Every neuron or thought connects, on average, to 50,000 other neurons. This is the vast connection of "the magic trees of the mind."

The Bible reveals that our brains' neuroelectrical pathways become powered by the BM-Field of the Holy Spirit when we align our minds with Christ. Every neuron, every thought we have is a flowing energy in the mind. Thoughts are invisible energy. As we transmit those same energies from the mind of Christ by meditating and praying. We then become partakers of the tree of life. In Genesis 2 and Revelation 22, the tree of life has a river constantly flowing beside it. In both verses, the Holy Spirit represents the river that flows from the throne of God and from the Lamb into the human spirit.

The same structural majesty of the Holy Spirit flowing from the throne of God and the Lamb hides the secrets behind the genesis of the mind of Christ, which are immediately accessible to every believer and elusive to every unbeliever. The flow of the Holy Spirit is the architectural principle of the mind of Christ. Through the flow, the Spirit brings all things to our remembrance. As Acts 8: 18-20 tells us, we receive knowledge and wisdom from Christ into our minds, so that we can master things without effort, and yet the exact opposite might be true for the unbeliever.

The Buffer Zone

The nonlocal BM-Field of the Holy Spirit represents your spiritual sight, or your mind's eye. No miracle has ever occurred through the ministry of Christ that has not been observed or meditated through the BM-Field of the Holy Spirit.

According to quantum physicists, without the interference of an observer or mediator, the quantum field would remain an invisible, nonlocal zone of energy. When we pray or perform miracles, our human spirit expands to a macroscopic scale, to observe the size of the entire universe and all spiritual possibilities—healing, provision, divine perspective, and so on. We see all possibilities within the BM-Field of the Holy Spirit. And we must bring those heavenly resources into our physical realm.

Everything we observe through this field will collapse into matter as we speak in faith. As Hebrews 11:3 explains, "By faith we understand that the *worlds* were framed by the word of God, so that the things which are seen [visible] were not made of things which are visible."

There are two worlds, and the nonlocal BM-Field is the buffer zone for your spiritual sight. Without the observer, these energy waves would not materialize. This is known as the *observer effect*. These fields illustrate the divine intervention of the mind of God in all its complexity to reveal His glory on the earth.

> Nothing in all creation is hidden from God. Everything
> is naked and exposed before his eyes, and he is the one to

whom we are accountable.
—Hebrews 4:13

This revelation is powerfully substantiated by the discoveries of the BM-Field of the Holy Spirit. "Who has known the mind of the Lord that he may instruct Him? But He has revealed some of the deepest mysteries to us by His Spirit" (1 Corinthians 2:16). When we probe the depths of His mind, God permits us to take a glimpse into some of the deepest mysteries of creation. We know this is how Jesus lived. He said, "I do nothing on my own initiative. I only do what I see the Father do and say what I hear Him say" (John 5:19). Let's look at how this revelation relates to neuroelectrical pathways.

The Effect of Words

Neurophysicist Eric Kandel won a Nobel Prize for his work on memory. He studied how our thoughts and emotions essentially affect our DNA, turning genes on and off and changing the structure of the neurons in our brains. His work shows that we are capable of changing the structure and function of our brains. Remember those neuroelectrical pathways? We can make them stronger with good, positive thoughts and emotions, and we can make them brittle and weak with negative ones.

God's will, mind, and heart extend over the entire universe. His BM-field is all-encompassing. Our minds and hearts are wired to think like God and to love like God. As Jesus said, "When He, the Spirit of Truth has come, he will guide you into all truth" (John 16:13). The Holy Spirit guides us into all truth; it's our job to ensure the BM-Field of our mind is positive and pure, so we can accept the truth.

Without our relationality to the BM-Field of the Holy Spirit, our finite mind is void of reality. Famed missionary Dr. Rolland Baker, who heads up Iris Global, puts it this way: "Sin to my mind is a rejection of relationality, and some of the most brilliant minds in history have refused the logic of relationality with God."

People who have rejected a relationality to the BM-Field of the Holy Spirit have a reprobate mind. A *reprobate mind* is a morally corrupt mind, one whose thoughts are unprincipled by the Word and depraved of the truth. By implication, the thoughts of a reprobate mind are worthless or wicked and devoid of the Holy Spirit.

> And even as they did not like to retain God in *their* knowledge, God gave them over to a debased mind, to do those things which are not fitting; being filled with all unrighteousness, sexual immorality, wickedness, covetousness, maliciousness; full of envy, murder, strife, deceit, evil-mindedness; *they are* whisperers, backbiters, haters of God, violent, proud, boasters, inventors of evil things, disobedient to parents, undiscerning, untrustworthy, unloving, unforgiving, unmerciful; who, knowing the righteous judgment of God, that those who practice such things are deserving of death, not only do the same but also approve of those who practice them.
> —Romans 1:28-32

Praying Amiss

When we think of the mind of Christ resonating through our neurons, it starts to make sense that people without the mind of Christ can pray amiss. Neuroelectrical pathways are a conduit for the mind of Christ, and the Holy Spirit does not have access to the carnal mind. Therefore, a person who prays amiss will likely continue to pray amiss. If you do not have the mind of Christ, your carnal mind is hostile toward God, so you are not praying with the mind of Christ. Consider the following:

> From where do wars and fighting's among you come? Is it not from this, from your lusts which war in your members? You desire, and do not have. You murder, and are jealous, and cannot obtain. You fight and war, yet you have not because you ask not. You ask and receive not, because you ask amiss, that you may spend it upon your lusts.
> —James 4:3

Here, James is explaining that the sinful mind or carnal mind is hostile toward God. It hates the very things of God and cannot please God because it is impossible to walk according to His ways while following the desires of the flesh. Lust is a temptation and an evil that overcomes the human spirit. In lust, your prayers resonate from carnal desires, and God is not going to answer any request that comes from the carnal mind. Many people experience failure while praying because they are not praying sincerely from the heart. They are praying from the sinful lust and desires of the flesh, and as we discussed, the carnal mind cannot capture the mind of Christ.

> And I, brethren, could not speak unto you as unto spiritual, but as unto carnal, *even* as unto babes in Christ. I have fed you with milk, and not with meat: for hitherto ye were not able *to bear it*, neither yet now are ye able. For ye are yet carnal: for whereas *there is* among you envying, and strife, and divisions, are ye not carnal, and walk as men? For while one saith, I am of Paul; and another, I *am* of Apollos; are ye not carnal? Who then is Paul, and who *is* Apollos, but ministers by whom ye believed, even as the Lord gave to every man?
> —1 Corinthians 3:1-5

The above scripture shows that the carnal mind works against the BM-Field of the Holy Spirit. The carnal mind does not allow the good thoughts to charge the neuroelectrical pathways or grow in goodness like the tree of life. Because neurological pathways are a conduit for the mind of Christ, we need to focus on the purpose of prayer, which is to bring the kingdom of God to us.

We may harbor these carnal thoughts, and we may not even realize that we have them until we pray. For example, a man may hate his neighbor and yet not discover his resentment until he prays. Your prayers will instantly bring out whatever you have spent countless hours observing.

Romans 8:7 states, "Because the carnal mind is enmity against God: for it is not subject to the law of God, neither indeed can be." This scripture is telling us, essentially, that the prayers of the carnal mind lack a true and acceptable purpose because they focus on the things of the flesh. Led by the carnal mind, Adam moved away from the mind of God and stepped into the realm of the flesh. As Paul said in Galatian 3:3, "Are you so foolish, having begun in the Spirit, are you now being made perfect by the flesh?"

He who minds the things of the flesh will pray amiss. However, he who minds the things of the Holy Spirit, will not pray amiss. He takes pleasure in observing thoughts from the mind of Christ, meditating upon them, and conversing about them. He reads about them in the Word of God, and he frequently prays to God for he is in relationality to the BM-Field of the Holy Spirit. He is alive in Christ, awake to the highest functions, and filled with witty ideas, dreams, visions, and innovations. He is pushing humanity forward with the mind of Christ, growing those good and pure branches in his neuroelectrical pathways.

The primary reason for a believer to have the mind of Christ can be summed up in first few words of The Lord's Prayer: "Our Father in heaven, hallowed be your name. Your kingdom come, your will be done, on earth as it is in heaven."

Prayer Purpose vs. Prayer Strategies

Praying amiss will also happen when you confuse the *purpose* of prayer with the *strategies* of prayer. Remember that your strategy will change based on your circumstances, but your purpose in your prayer should remain constant.

Purpose is defined as the reason for which something is done or created, or for which something exists. There is a powerful truth in the statement: "We misuse a thing when we don't understand its purpose."

Your purpose in prayer should remain constant: "Rejoice always, pray without ceasing, give thanks in all circumstances; for this is the will (purpose) of God in Christ Jesus for you" (Thessalonians 5:16). By living in union always with God, we experience wholeness and fulfillment, giving ourselves to the very purpose of our creation. Your strategies for prayer will change based on your circumstances (warfare, praise, uncertainty, agony, pain). So, depending on what's going on in your life, your strategies may change, but your purpose should always be to bring the will and kingdom of God to earth.

Christ's mind was always on obeying His Father's will and manifesting the kingdom of God on Earth. So, just like your mind is a conduit for the Father's will, your neuroelectrical pathways are a conduit for the mind of Christ. If you stay the course, your purpose in prayer will never change. If you keep to that purpose, you will not be praying with a carnal mind. You will not pray amiss.

A True Transformation

As we discussed, Bioglory Morphic Resonance has a profound implication on prayer-encoding Life-Light, which can map onto the human genome to store and reproduce the exact blueprint image of Christ in bodily form.

Therefore, all that Christ did in the flesh while on earth—healing the sick, raising the dead, taking the keys of death, loving and caring for others, and denying his own will to do the will of God—continue to affect the morphic field and enable others to walk the same path.

Continue to walk in the Spirit. Continue in prayer. Although you may not see it immediately, you are being transfigured through prayer! You are literally becoming like Christ. Your neuroelectrical pathways are branching into a tree of life to manifest God's will on earth.

It is possible to engage so deeply with the Lord of glory in prayer that we experience a total transformation of our DNA. Through science and the Bible, we see that metamorphosis is entirely possible. When we pray, something always shifts in the quantum and material worlds.

I have made a commitment to spend countless hours unveiled in the presence of God; and soon my cells will have absorbed His Bioglory to a degree that the atoms in my body will began to break down my flesh to the smallest particle—and one day only the *exact blueprint* of the image of Christ will remain.

Through prayer, I am achieving a higher level of glory. Already, I have changed. I am not the same. My life has changed. My situation has changed. My attitude has changed. My mindset has changed. Everything around me has changed for my benefit. The Father chose me, just as he chose you, at the beginning of creation "to be conformed to the image of his Son, in order that he might be the firstborn among many" (Romans 8:29). He has already begun this good work on the inside of me, and He will faithfully "bring it to completion at the day of Jesus Christ" (Philippians 1:6).

If you still have doubts about the power of praying with the language of glory, read Hebrews 11:5:

> By faith Enoch was taken from this life, so that he did not experience death: "He could not be found, because God had taken him away." For before he was taken, he was commended as one who pleased God.

Enoch was in habitual prayer and fellowship with God, until his physical presence was de-materialized. Because of faith, he never saw death. Something greater and faster than time was at work on the inside of Enoch's cells.

The Transformative Power of Prayer
God revealed the fullness of Himself to us through Christ. Consider the following scriptures:

> That which was from the beginning, which we have heard, which we have seen with our eyes, which we have looked upon, and our hands have handled, concerning the Word of life—the life was manifested, and we have seen, and bear witness, and declare to you that eternal life which was with the Father and was manifested to us—that which we have seen and heard we declare to you, that you also may have fellowship with us; and truly our fellowship *is* with the Father and with His Son Jesus Christ. And these things we write to you that your joy may be full.
> —John 1:1

But blessed are your eyes because they see, and your ears because they hear. For truly I tell you, many prophets and righteous people longed to see what you see but did not see it, and to hear what you hear but did not hear it. The prophets and righteous saw afar off the glory of the kingdom that we are beholding in the latter days.

—Matthew 16: 13-19

We have the morphogenetic field of Christ resonating within us, which can help us achieve a greater level of Bioglory. I can mentally affirm that through prayer, through the transformation of the neuroelectrical pathways of my mind, something greater and more powerful is now at work on the inside of me. Through prayer, my mind is becoming more like the mind of Christ.

Through prayer, your mind can become more like the mind of Christ as well. The Holy Spirit will flow like a river of living water, bringing forth good thoughts, inspirations, and perceptions. I encourage you to spend time with God in prayer and meditation. It will improve your neuroelectrical pathways and you will become like a tree that bares good fruit.

CHAPTER 7
EPIGENETICS AND SPIRITUAL TRUTH

For then I will restore to the peoples a pure language, That they all may call on the name of the LORD, To serve Him with one accord.

— Zephaniah 3:9

God has promised to restore a "pure language" to His people. I posit that this pure language is a genetic language that will be encoded within our DNA, which will bring us to a place of prayer, revelation, and intimacy with Him through Christ Jesus.

Previously, we've documented that DNA is the same for all living organisms and that what makes us different is the encoded information or language within that DNA. That encoded information makes us who we are today. To function correctly, each of your cells depends on duplicating the correct language. When the correct genetic code is not replicated, a gene mutation occurs within the DNA.

An organism's DNA affects how it looks, how it grows, and how it behaves. DNA also affects an organism's physiology. Therefore, a change in an organism's genetic code can cause changes in all aspects of its life—both physical and spiritual.

Epigenetics refers to DNA modifications that affect gene activity without changing the DNA sequence. Epigenetic signals can switch the gene on or off. This is important for our purposes because through epigenetics, our words and thoughts can modify our DNA. Remember we talked about how our thoughts, words, and prayers can affect us on a molecular level and how these effects can be passed down for generations? This is epigenetics at work. In this chapter, we will explore how epigenetics can reveal spiritual truth.

A DNA Change Took Place

I believe that sin is a lethal fixed mutation that switched off divine gene expressions within Adam and Eve, resulting in a degenerative metamorphosis. The DNA change that occurred is apparent in Genesis 2:22-25, when they found themselves naked and hiding from the presence of God.

The human genome has continued to degenerate ever since, but what must it have been like before mankind sinned? Would Adam and Eve literally have exuded, duplicated, and reflected God's Bioglory? If God is Life and Light, and He breathed Himself into Adam, wouldn't every cell in Adam's body have encoded Life and Light? And then what about the release from inside of Adam's DNA? Wouldn't he have exhaled Bioglory that connected him to every conceivable aspect of the universe? Wouldn't this essence of Bioglory Life-Light engulf Adam and all of creation like a garment, fashioned after the glory and splendor of the Father? Yes. The Bioglory of God would have infused Adam's body and all that he was given dominion over.

When the lethal genetic code of sin was encoded into Adam, the Bioglory genetic codes were switched off in Adam's genes. Adam and Eve were deceived by Satan because he came to them as an angel of light, and his genetic codes looks similar to the original Bioglory codes. Satan is crafty, and he knew that a replica must appear to be the real thing. Therefore, he came to them with information, persuasive ideas that appeared to be the truth of God's Word.

> Now the serpent was more crafty than any of the wild animals the LORD God had made. He said to the woman, "Did God really say, 'You must not eat from any tree in the garden'?"
> —Genesis 3:1

Here, for the first time, Satan has introduced Eve to his fruit, his mutated genetic code in the form of a word sequence. Eve's DNA had already encoded the original Word sequence from God (Genesis 2:16-17). Therefore, her DNA immediately recognized the error in the genetic code—the word sequence that came from the serpent that could potentially modify her DNA. Let's review the next scripture:

> Then the woman said to the serpent, "We may eat fruit from
> the trees in the garden, but God did say, 'You must not eat
> fruit from the tree that is in the middle of the garden, and
> you must not touch it, or you will die.'"
> —Genesis 3:2

Eve's verbal utterance of the original Word sequence that God spoke to her and Adam signifies that her DNA encoded the Word. At this point, her DNA was not damaged. Satan's mutated genetic code had no point of entry.

But Satan knew that gaining an access point into the original genetic code of the DNA must take place at the right time—so he repeated the word sequence. But this time, he added the mutated codes:

> "You will not certainly die," the serpent said to the woman.
> "For God knows that when you eat from it your eyes will be
> opened, and you will be like God, knowing good and evil."
> —Genesis 3:4

All words resonate within the chambers of your spirit. They carry information in and out of your being. If, for example, Eve took the words spoken by the serpent, "You will not certainly die" and sounded them silently within the spirit of her mind, this would create an electromagnetic pulse that could be heard by every cell throughout her entire body. If she then would have imagined the feeling that the serpent's words engendered, she could have sent an even more powerful signal into the inner recesses of her spirit.

> When the woman saw that the fruit of the tree was good for
> food and pleasing to the eye, and also desirable for gaining

wisdom, she took some and ate it. She also gave some to her
husband, who was with her, and he ate it.
—Genesis 3:6

When Eve *saw*, she had a mental image of the fruit, which appealed
to the lust of her eyes. When Eve *desired*, she had a feeling of longing,
which appealed to the lust of her flesh. These lusts then permeated her
body, and she acted accordingly.

Controlling the Switches

The serpent knew that DNA is so incredibly sensitive that it hears everything (thoughts, words, and sound frequencies) and responds accordingly.
And by using this knowledge to his advantage, he was able to gain a point
of entry to insert his genetic mutations. In this way, the serpent gained
access to the human genome, and through epigenetics, he altered Eve's
DNA and that of humanity for generations to come.

Even a slight reception of the serpent's error codes in your DNA
will create enough of an epigenetic change to trigger your gene switches.
"Then when lust hath conceived, it bringeth forth sin: and sin, when it is
finished, bringeth forth death. Do not err, my beloved brethren" (James
1:15-16). "Do not give the devil a foothold or point of entry" (Ephesians
4:27). Footholds are points of entry for spiritual darkness to create fixed
mutations that are protected by strongholds within the mind.

Fixed mutations, like spiritual strongholds, are dwelling places for
darkness. A stronghold is a fortress that is heavily defended from intrud-
ers. In like manner, a fixed mutation is the error-ridden genetic code
within the DNA sequence that is heavily defended by the enemy.

The "memory" of DNA does not extend beyond the current state. If
the original genetic codes are altered and not corrected, the DNA will
reproduce the fixed mutation. When a genetic code is fixed, this mutated
sequence is automatically passed on to every succeeding generation,
ready to predispose children before they are even conceived. In Psalms
51:5, David said, "Behold, I was brought forth in iniquity, And in sin my
mother conceived me."

Paul understood that the fixed mutations that were passed down from the first Adam could only be destroyed by the second Adam, who is a life-giving spirit:

> And so it is written, The first man Adam was made a living soul; the last Adam *was made* a quickening spirit. Howbeit that *was* not first which is spiritual, but that which is natural; and afterward that which is spiritual. The first man *is* of the earth, earthy: the second man *is* the Lord from heaven.
> —1 Corinthians 15: 45-47

> How much more will the blood of Christ, who through the eternal Spirit offered Himself unblemished to God, purify our consciences from works of death, so that we may serve the living God!
> —Hebrew 9:14

Now that we've been given Christ, who is the genetic code of Life-Light, our goal is to encode His genetic language and make our lives better by erasing the genetic mutations of sin and death.

Erasing the Mutation

Since the discovery of DNA and the mapping of the human genome, a huge amount of energy has gone into trying to identify and locate genes that seem to have specific links to certain mutations, diseases, and illnesses. It has been proven that mutations, diseases, and illnesses are caused by epigenetics, linked to genes that respond to thoughts, words, sounds, and frequencies, and these have their own codes that could potentially enhance, help erase, or correct them.

For example, knowing your genetic predisposition to heart disease may well end up being a double-edge sword because this worrying knowledge could lead to stress, which increases the possibilities of contracting heart disease. On the other hand, feelings of optimism will release frequencies to the gene, which could help correct your DNA.

However, there is an even higher truth hidden here. God created the human genome system, and He knows the locations of every gene, so He is able to reprogram, correct, and totally erase any genetic code. All genetic codes can be transcended and are in fact designed to be transcended by the Creator.

> The LORD said to Satan, "The LORD rebuke you, Satan! ... There are seven eyes on that one stone, and I will engrave an inscription on it," says the LORD Almighty, "and I will remove the sin of this land in a single day."
> —Zechariah 3

Through genetic coding, The Lord can totally erase fixed mutations (sin and iniquity) to heal disease or end illness for an entire nation or bloodline. Consider some other examples:

> The Lord says, "I, yes I, am He who blots out your transgressions for My own sake and remembers your sins no more."
> —Isaiah 43:25

> He sent His word and healed them, And delivered *them* from their destructions.
> —Psalm 101:2

> He will again have compassion on us; He will vanquish our iniquities. You will cast out all our sins into the depths of the sea.
> —Micah 7:19

> Then shall thy light break forth as the morning, and thine health shall spring forth speedily: and thy righteousness shall go before thee; the glory of the LORD shall be thy reward.
> —Isaiah 58:8

If my people, which are called by my name, shall humble themselves, and pray, and seek my face, and turn from their wicked ways; then will I hear from heaven, and will forgive their sin, and will heal their land.

—2 Chronicles 7:14

Replacing Codes

Multiple scientific studies have revealed that fixed mutations can be erased and replaced by good genetic codes. The *Nature Biotechnology* newsletter describes these findings in layman's terms, and it clearly explains that the gene-editing technique CRISPR can reverse disease symptoms in living animals:

> CRISPR, which offers an easy way to snip out mutated DNA and replace it with the correct sequence, holds potential for treating many genetic disorders, according to the research team,. . . about half of the 32,000 known point mutations that are linked to diseases are down to bases that ought to be G instead being A and their corresponding pair being a T instead of a C. They can fix these errors in a process known as "base editing," turning A bases back to G and T bases back to C using a modified version of the gene editing tool Crispr–Cas9.
>
> —Nature.com

Praying with the language of glory helps restore what has been diseased or broken within the human genome, and that kind of restoration goes beyond the curing of our physical illnesses or the epigenetic editing of fixed mutations.

Adam's fixed mutation of sin encoded the information or genetic codes for death to rule over all living organisms (human, animal, plant) as a predisposition. Adam and Eve underwent a degenerative metamorphosis, which God reprogrammed by restoring Bioglory, the Life-Light of His Word, to the human genome. Change will come to our DNA through the encoding of God's Bioglory.

We documented that DNA acts as a superconductor and that it can exponentially increase the frequency of Bioglory Life-Light passing in, out, and around your body. This influx can lead to a complete transformation of the fabric of your being. This process allowed the garment of God's Bioglory to cover Adam and Eve, and you can also receive this garment through the power of prayer. DNA can weave Life-Light within itself, which is its true hidden role within your body.

Always remember, prayer is not a complicated thing. Your mind may try and make it complicated, but really, it is as natural as breathing. You can pray while you go about your everyday life—working, looking after the children, cleaning the dishes, relaxing. God will even visit you in your sleep. Once you've established prayer patterns, your DNA will continue to encode and weave Life-Light within you while your mind is resting. The Bioglory garment that covers you is a direct implication of the Life-Light that is being encoding inside your DNA. If you find yourself naked and without a Bioglory garment, it is because you do not have a prayerful life.

Nakedness

I propose that "nakedness" was the first degenerative metamorphosis that took place after the sin mutation was fixed within Adam and Eve. Before Adam and Eve sinned, Genesis 2:25 tells us, "And they were both naked, the man and his wife, and were not ashamed." After they sinned, that changed:

> God called unto Adam, and said unto him, "Where art thou?"
> And he said, "I heard thy voice in the garden, and I was afraid, because I was naked; and I hid myself."
> —Genesis 3:9

Adam and Eve knew something was different, so they hid themselves out of fear. Is it possible that the word "naked" in Genesis 3 is describing the change in their gene expression? For a fuller understanding, Let's take a peek at Lucifer's garment before he was cast out of Heaven:

Son of man, take up a lament for the king of Tyre and tell him that this is what the Lord GOD says: "You were the seal of perfection, full of wisdom and perfect in beauty. You were in Eden, the garden of God. Every kind of precious stone adorned you: ruby, topaz, and diamond, beryl, onyx, and jasper, sapphire, turquoise, and emerald. Your mountings and settings were crafted in gold, prepared on the day of your creation. You were anointed as a guardian cherub, for I had ordained you. You were on the holy mountain of God; you walked among the fiery stones. From the day you were created you were blameless in your ways until wickedness was found in you."

—Ezekiel 28: 12-15

Lucifer was created with genes that were designed to express the Bioglory of God as a garment which covered his body; however, after he was cast down, Jesus said:

"How you are fallen from heaven, O Lucifer, son of the morning! *How* you are cut down to the ground, You who weakened the nations!"

—Isaiah 14:12

In the Garden, as a serpent, Lucifer was naked—stripped of his Bioglory garment. Now it is plain to see why Lucifer wanted Adam to be out of fellowship with God. Lucifer lost his Bioglory garment when he was cast out of God's presence, and he wanted Adam and Eve to suffer the same fate, a goal he accomplished through epigenetics.

Even today, Lucifer appears as an angel of light. So it is no surprise that his servants also disguise themselves as "servants of righteousness." This is the most important bit of information all believers need to remember concerning the nature of Lucifer and his work: he has a counterfeit Bioglory garment and so do his servants. This garment is not the real thing! His corrupt doctrines, habits, and values will lead to eternal death and destruction. For he is the "father of lies" (John 8:44).

In our nakedness, we will always do two things: run from prayer and fellowship with God and cover our nakedness with a counterfeit Bioglory garment.

Running from prayer and fellowship with God has become the hallmark of human nature ever since the fall. When mankind lost fellowship with God, they lost His Bioglory garment along with the protection that garment provided. The carnal man has been attempting to replace the fellowship of God with religion for centuries. Religion is doctrine without the transformational experience of the Life-Light of God. Adam and Eve lost their focus, and their attention drifted from God. In that moment, they desired the fruit more than they desired God. When they fell out of fellowship with Him, they tried to cover themselves the best they could, using their carnal senses to guide them.

Earlier we talked about the carnal mind, and how being of a carnal mind is an enmity against God: for the mind is not subject to the law of God, neither indeed can be (Romans 8:7). When Adam and Eve lost fellowship with God, their carnal minds thought they needed to hide and cover themselves before the Lord.

The Seat of Procreation

The carnal man is determined to make a lookalike garment of Bioglory to cover the nakedness and shame that the original sin created. If we carefully examine Adam and Eve's actions in the garden, we witness a cunning attempt to replace the original garment of Biology Life-Light with a manmade counterfeit covering. Genesis 3:7 tells us that "they made themselves aprons of fig leaves to cover their loins." Adam and Eve used the leaves of the fig tree to sew garments (aprons) in an effort to replace the Bioglory garments they had lost.

The loin is the physical and spiritual seat of procreation. If we are covering this area with something other than God's Bioglory garment, then we are not reproducing from Him. Consider the following passages:

Therefore gird up the loins of your mind, be sober, and rest your hope fully upon the grace that is to be brought to you at the revelation of Jesus Christ.
—1 Peter 1:13-16

Let your loins be girded about, and your lights burning; And ye yourselves like unto men that wait for their lord, when he will return from the wedding; that when he cometh and knocketh, they may open unto him immediately. Blessed are those servants, whom the lord when he comes shall find watching.
—Luke 12:31-48

But put on the Lord Jesus Christ, and make no provision for the flesh, to fulfill its lusts.
—Romans 13:14

Stand therefore, having your loins girt about with truth.
—Ephesians 6:14

The mind is also a place of procreation. It is where you produce your thoughts. Their minds were no longer covered by God's garment of Bio-glory, so they procreated thoughts and emotions from a carnal place.

Remember that the carnal mind is naturally predisposed to create its own laws and ways of doing things. So if you listen to your carnal mind, you cannot have thoughts, emotions, or behaviors that are in communion with the mind of Christ. Adam and Eve found themselves naked, so they created their own manmade coverings. Similarly, you might fool yourself into thinking that certain rituals or actions are pleasing to God.

As soon as Adam and Eve partook of the fruit, their hearts were changed. Satan altered their DNA, and sin immediately marred the quality of the couple's relationship with God. When fear entered their hearts, their separation from God was complete.

We have all expressed the fear, guilt, and shame sparked by the genetic codes of sin. For a short period of time in my life, I ran from fellowship with God because of my own sin and shame. I knew the Word of God, and I was able to quote the scriptures, but my heart was not humble before the Lord. I only feared the Lord because of my own self-righteousness, not because I wanted to humble myself before Him to be changed.

Self-righteousness, to my understanding, is a form of fear that can keep us from turning to God to have our sinful natures addressed. Because we identify as Christians, we sometimes feel we are "good enough," and this line of thinking reveals a level of arrogance and pride that is often rooted in religion.

As human beings, we should not be covering our own nakedness or our own mistakes. We should submit ourselves in prayer to God and allow Him to cover us with His garment of Bioglory. Any time we cover ourselves or our shortcomings, we produce things from our own carnal minds—things that are outside of the will of God. But when we allow ourselves to come into His presence with our mistakes and our nakedness, and be real about who we are, that's when God is able to heal, deliver, and set us free. He is able to correct our DNA. Through epigenetics, He is able to transform us.

Jesus and the Fig Tree

In Matthew 21, we learn about the metaphorical unproductive fig tree, which mantled the loins of the first Adam and is now being dismantled by the second Adam (1 Corinthians 15:45-49). Jesus destroyed this fig tree from the root by the power of His spoken Word. The tree withered from the inside out. Jesus spoke directly to the subatomic particles that formed the fig tree, and the elements at the quantum realm obeyed His words.

Jesus understood the subatomic realm because He wrote the master codes of the universe. Jesus is the true garment of Life-Light. He is the Word made flesh, and His loin is girded with truth to curse the counterfeit fig tree garment.

In John 12:1-19, Jesus travels to Bethany. The word *Bethany* translates as "house of affliction" or "house of figs." Prophetically, the fig tree garment brings affliction and barrenness not only to the human spirit but also to the physical body. When Jesus came in contact with the fig tree, it appeared to have fruits, but its branches were barren. "Therefore He cursed it from the root, saying, 'Let no fruit grow on you henceforward forever.' And immediately, the fig tree withered away" (Matthew 21:19).

Like the serpent, the fig tree is deceiving and crafty. It appears to be fruitful but it doesn't produce fruit. The fig tree was cursed because it lied. The fig tree is a metaphor for the carnal man. The carnal mind will always be deceptive, even while posing as something good and fruitful.

Abstract Prayers

Some of us are living lies. Some of us simply don't understand that God wants us to have abundant spiritual lives. So, we focus all of our time and energy on producing different kinds of fruit: success, wealth, and fame. Achieving these goals gives us a false sense of accomplishment, of a life well-lived. But God wants us to experience spiritual truth. He wants us to achieve a spiritual life.

The carnal man is constantly geared toward replacing a genuine relationship with God with actions, rituals, and traditions. Many have become so faithless that the notion of living a lifestyle in alignment with God has become an abstraction. In addition, most people have forgotten that they can indeed be clothed with the Bioglory garment.

We have been practicing abstract prayers in church services. We have taken away what I like to call the "faith-energy" characteristics of prayer, the same faith-energy that Christ used to curse the fig tree from the root. This faith-energy is the substance of things hoped for and the evidence of things not seen.

Many have reduced prayer to a ten-second religious portion of our worship services. Many are no longer looking for the substance of the Word to bring forth that which we have hoped for. The idea that one can spend time in God's presence until he or she literally illuminates with the power and glory of God is relegated to antiquity, along with the Biblical accounts of men who accomplished that feat while praying.

Remember that science has already proven that DNA has the ability to encode and duplicate light. DNA can store the light energy needed to change molecules. Therefore, I believe the church can reference the transfiguration of Christ and teach church members to seek his Life-Light through prayer. Scientifically, the transfiguration of Christ is possible. If we, in church, start praying again with "faith-energy" characteristics of prayer, we can become transformed through God's Life-Light. God can change our situations and circumstances; He can set us free.

If divine illumination happened to Enoch, Moses, Stephen, Jesus, and also to the apostles on the Day of Pentecost, what are we missing in our modern-day church? Jesus wants us to be like Him. Ephesians 5:27 says that He wants the church to be a "glorious church." The New International Version translates this as a "radiant church." The Good News Translation says that His church will be a church "in all of its beauty." All of these descriptions point to a church that is consumed with God's Bioglory. I am fully persuaded that those who become the "sons of God" will be those who spend ample time with God in prayer.

As an individual created in the image and likeness of God, your DNA is multifaceted. You are not just three different parts—spirit, soul, and body. You are a complex organism made up of trillions of cells, each with their own structures and functions.

The Structure of DNA

DNA is crystalline and hexagonal in shape, so it can absorb positive or negative energies from your thoughts. Thoughts are silent words, and words are multidimensional elements that can affect both the physical and spiritual worlds at the same time. Thus, when you think negative thoughts and have feelings of fear, anger, and frustration, the hexagonal shape of the DNA will tighten up, reshaping the folds and dimensions and triggering codes to switch off multiple DNA genes, thereby reducing the quality of your bio-photon emissions and your external body expressions. You will begin to flow and function from a lower frequency that expresses unstable emotions.

But here's the good news: whenever your DNA hexagonal shape is distorted, you can reconstruct the shape by simply encoding different information. By encoding Bioglory Life-Light, focusing on positive energy, and doing the things that are pure, lovely, and true, your DNA sequence will change again. It will loosen up to allow energy and information to flow, switching on genes that were deactivated and restoring everything back to its original healthy state.

We previously documented that sin is a lethal mutation that entered the human genome (DNA) through Adam. Mutations can switch genes on or off. When a mutation occurs, it can substantially change the way an organism is built, the way it looks or thinks. Your spirit, soul, and body can be affected by a mutation in your genetic code.

When a mutation occurs in a gene, the impact is like a conductor instructing all members of an orchestra to play a different tune. The result is disharmony. Consider this scripture:

> For I delight in the law of God, in my inner being, but I see in my members another law waging war against the law of my mind and making me captive to the law of sin that dwells in my members.
> —Romans 7:23

Mutations within your inner being can wage war against the rest of your body, soul, and mind by reprogramming the behavior of every cell connected to that particular DNA switch. This mutated genetic code then becomes a new law in the body.

Taking this idea further, Romans 3:23 explains that "everyone has sinned; we all have fallen short of God's glorious standard." Sin is a genetic law or mutation that started in Adam's DNA.

Epigenetic scientists have proven that DNA mutations are real, and they can affect the physical and chemical flows in our bodies. Whatever genetic mutation your DNA is encoding becomes the reality of your identity. Every day, scientists are discovering new and amazing things about DNA. Even so, some of the greatest scientific insights in the field may lie in our future.

But you don't have to wait for science to prove anything to you about your divine identity. You already have free access to the laboratory of your DNA because you have been given free will. Your DNA is simply waiting for you to input the right code or information. Once it has received the right instruction, it will run the new program and build you a new mind, a new life, and a new reality.

But if your DNA receives the wrong codes or information, you will self-destruct. Your mind, life, and reality will deteriorate. The evidence is lying within the code or information that has been presented to you:

> And the LORD God commanded the man, saying, "Of every tree of the garden you may freely eat; but of the tree of the knowledge of good and evil you shall not eat, for in the day that you eat of it you shall surely die."
> —Genesis 2:16-17

God gave Adam information that there were two different trees and both had fruits with seeds in them. The seeds represent the genetic codes or information that were embedded within the fruit. Every seed has its very own genetic code, and you, like Adam, have the power to choose which genetic code will modify and shape your identity.

> Today I have given you the choice between life and death, between blessings and curses. Now I call on heaven and earth to witness the choice you make. Oh, that you would choose life, so that you and your descendants might live!
> —Deuteronomy 30:19

The Law of the Harvest

Your words and actions create genetic modifications that carry into your future; therefore, it is God's desire that you choose life—that both you and your seed may live. Remember that DNA has the ability to encode, store, and recall information that can impact generations to come (Exodus 20:5, Exodus 34:7, Numbers 14:18). Your mental perceptions will influence your choice. The moment God created man, Adam entered into the realm of creativity and responsiveness. He was held accountable for making intelligent choices within his environment. So, put yourself in nurturing environments that are pleasing to God.

Every single cell in your body is sending and receiving information from your environment through countless molecular antennae embedded into the cellular membrane. This cellular outer layer interprets the environmental signals and relays the information to the DNA within your cells. Your DNA then responds by activating the necessary machinery within the cell.

If the outer layer of the cell comes in contact with the wrong information, or codes, the DNA within the nucleus of the cell will either open or close access by switching genes on and off. This process occurs in all sixty trillion cells all the time, day and night. As long as you are alive, your cells will be sending and receiving codes or information from your environment.

What all of this means is that you can never be the victim of your DNA, but your DNA can be the victim of your intentional choices. Every thought you think or receive, every word you speak or hear, every feeling you send out or take in, every action you complete or submit to directly programs your genes and therefore your reality. Your chosen environment is conditioning you at a quantum level.

DNA will reproduce according to the information encoded within it. In like manner, a seed will only reproduce according to its unique genetic codes (Matthew: 12:33). A corn seed will only produce corn. A blackberry seed will only produce a blackberry bush. And an avocado seed will only produce an avocado tree. The Law of the Harvest is a principle of encoded information; this Law states that each fruit tree can only yield fruit after its kind, in accordance to the seed that was planted.

Blooded Priestly Garment

You have seen how your genes hold the blueprint of your nature. We've also studied how every thought, word, feeling, and action—all the information that you send and receive—will generate subtle electromagnetic signals that have a profound effect on your DNA. And epigenetics shows that those mutated genes can be passed on to the next generation.

Before the fall, Adam had been the high priest of the earth. There was no sin, only fellowship with God. After the fall, God immediately removed the fig tree garment that Adam and Eve used to cover their nakedness. God provided a *blooded priestly garment* in its place. "The LORD God made garments of [blooded] skin for Adam and his wife and clothed them" (Genesis 3:21). This gift was an act of prophetic implication, foreshadowing that God would restore Adam's blueprint as high priest and give His Bioglory garment back to mankind.

Following this same pattern, The Lamb of God, Christ, the Great High Priest over all creation, came in the fashion of Adam (in the same flesh but without the genetic mutation of sin within His DNA). Christ, the second Adam, came to earth with the original blueprint of the Father encoded within His DNA. Thus, Christ was not naked and ashamed. He was clothed from the womb with the Bioglory of God, and we were transformed.

Jesus is the Great High Priest forever. He never takes a break from prayer or goes to sleep on His throne. The genetic codes of death, hell, and the grave will never prevent Him from executing His office as our High Priest.

Every born-again believer is clothed with the blooded priestly garment of Christ. Unlike the Levitical Priesthood, whose garments were crafted by men, the blooded priestly garment is encoded by the blood of the Lamb.

In the ancient temple, the Levitical priest would sprinkle blood around the altar and the entryway after every sacrifice. When priests performed these rituals in the temple, the blood from each sacrifice would cover or stain their garments. This covering of blood was symbolic of the blooded priestly garment that the coming Lamb of God would provide (Leviticus. 3:2).

We are living in a time when the blooded priestly garment is becoming manifested upon believers in ways that have not been seen since the days of the early church. As people seek the face of God and experientially cross over the threshold into the most holy place, they are beginning to understand the significance of being covered in the blood of Christ.

Cells are Bioglory Carriers

According to the Melchizedek priesthood, Christ has transitioned us from the Old Testament "Ark of the Covenant" to the New Testament "Temple of God." Through Christ, we have been given a more superior order than the Levitical Priesthood was given. Levitical priests served within a manmade temple. Today, we are partakers of the new covenant, a covenant that comes directly from God. Our cells are Bioglory carriers, ready and waiting to receive His Life-Light. Consider the following scriptures:

> "The glory of this latter temple shall be greater than the former," says the LORD of hosts. "And in this place I will give peace," says the LORD of hosts."
> —Haggai 2:9

> Forasmuch as ye are manifestly declared to be the epistle of Christ ministered by us, written not with ink, but with the Spirit of the living God; not in tables of stone, but in fleshy tables of the heart. And such trust have we through Christ to God-ward: Not that we are sufficient of ourselves to think any thing as of ourselves; but our sufficiency is of God; Who also hath made us able ministers of the new testament; not of the letter, but of the spirit: for the letter killeth, but the spirit giveth life. But if the ministration of death, written and engraven in stones, was glorious, so that the children of Israel could not stedfastly behold the face of Moses for the glory of his countenance; which glory was to be done away: How shall not the ministration of the spirit be rather glorious? For if the ministration of condemnation be glory, much more doth the ministration of righteousness exceed in glory.
> —2 Corinthians 3:9,10

I found some intriguing pieces of Biblical and scientific information regarding how and why our DNA has the ability to encode the blooded priestly garment, which now mantles our bodies, for we are the temples of the most holy God:

> But ye are a chosen generation, a royal priesthood, an holy nation, a peculiar people; that ye should shew forth the

praises of him who hath called you out of darkness into his
marvellous light
—1 Peter 2:9

Therefore, the blooded priestly garment is much greater than a powerful theological concept. It is a powerful reality in both the physical and
spiritual realms, for God, Himself has established it:

> And hast made us unto our God kings and priests: and we
> shall reign on the earth.
> —Revelations 5:10

> And from Jesus Christ, *who is* the faithful witness, *and* the
> first begotten of the dead, and the prince of the kings of the
> earth. Unto him that loved us, and washed us from our sins
> in his own blood, And hath made us kings and priests unto
> God and his Father; to him *be* glory and dominion forever
> and ever. Amen.
> —Revelations 1:5-6

Let's look at how DNA has similar qualities to the blooded priestly
garment. For one, DNA is crystalline and can absorb light and fluoresce.
The DNA within the cell is a hexagonal folded crystal structure, approximately ten atoms wide by six feet long. Biophysicists have also proven
that the physical bodies of human beings are composed of fluoresce liquid
(blood) crystals. A liquid crystal is a state of matter that has properties
between those of conventional liquids and those of crystalline solids. For
instance, a liquid crystal may flow like a liquid, but its DNA molecules
may be oriented in a crystalline way, where atoms and molecules are
arranged with a degree of order far exceeding any other substance.

Biophysicist Mae-Wan Ho explains this phenomenon in her article,
"Pursuing the Science of Global Coherence":

> Organisms are so dynamically coherent at the molecu-
> lar level that they appear to be crystalline . . . There is a

dynamic, liquid crystalline continuum of connective tissues and extracellular matrix linking directly into the equally liquid crystalline cytoplasm in the interior of every single cell in the body. This liquid crystallinity gives organisms their identity, characteristics . . . responsiveness . . . enabling the organism to function as a whole. The organism (human body) is cohort beyond our wildest dreams.

This discovery is even more exciting when we learn that the human body is not so different from the twelve crystalline minerals used in the ancient Levitical priesthood breastplate. As a matter of fact, our DNA is dynamically coherent to properties of the ancient Levitical priesthood breastplate.

TWELVE CRYSTALLINE STONES	DNA CRYSTALLINE HEXAGONS
Crystalline stones encode and store light energy in their molecules.	DNA is a crystalline molecule, a living stone that encodes and stores light energy.
God instructed Moses on how to design the breastplate with the twelve stones. "They fashioned the breastplate," which was "the work of a skilled craftsman."	DNA is an intelligent design. The origin of the genetic information needed to create the first cell must have came from an intelligent and skilled Creator.
The twelve stones represent the gene identity and the unique characteristics of the Twelve Tribes of Israel, which now includes all who have been born again of the Holy Spirit (Exodus 28:20-21, Matthew 21:43, Genesis 12:1-3, Galatians 3:8).	The DNA double helix encodes gene identity, which includes the physical, psychological, and social identity of the organism. All DNA is the same; the encoded information forms the differences between species. New information can transform the organism.

TWELVE CRYSTALLINE STONES	DNA CRYSTALLINE HEXAGONS
The book of Exodus in the Bible and the Torah describes how God himself instructed people to make a breastplate with twelve different gemstones for a high priest to use in prayer.	When we pray with unveiled faces, in the presence of God, our crystalline molecules encode and store His Life-Light information as a genetic language.
The twelve stones will resonate a fluorescent rainbow color, a banner of light in the most holy place. They are tied to the electromagnetic fields.	Fritz Albert Popp discovered that DNA cells light up with bio-photon emissions, creating a tiny laser light display. We have electromagnetic fields or auras of fluorescent light surrounding our bodies.
The twelve generational names engraved as a seal upon the stones contain a genetic language (Psalms 139:13-14). Also, the names of those who have been born again are engraved in the book of life, which is the actual Body (DNA) of the Lamb. "In Him we live and have our being." Only the Lamb has the DNA markers required to reveal the scroll in heaven. The Lamb's DNA is the volume of the book (Revelations 3:5; 20:12; Malachi 3:16). They are sealed with the name of God.	DNA's double helix is multifaceted and multi-generational, a genetic language that contains sophisticated syntax (codes, letters, sequences) that is encrypted. Your DNA is the volume of your book of life. It is 125 Billion miles long and has millions of trigger switches that can turn genes on and off. DNA can encode information through light to transform your physical/spiritual being and change genetic defects. DNA barcodes are for species identification.

TWELVE CRYSTALLINE STONES	DNA CRYSTALLINE HEXAGONS
The twelve stones also represent the twelve gateways to the holy city of God. "But I saw no temple in the city, because the Lord God Almighty and the Lamb are its temple." The wall of this holy city has twelve foundations, which are the named after Christ's twelve apostles. Combined, the apostles became the gateway on the Day of Pentecost through which the Holy Spirit and the tongues of fire (the genetic language of the Spirit) came to the church. Your body is the temple of the Lord.	The double helix is a gateway, a ladder, and a city of networking lights (bio-photons or Life-Light). Dr. Peter Gariaev discovered that DNA can absorb light and store its exact blueprint for more than thirty days.
The twelve-stone breastplate was also called the "breastplate of judgment" for the twelve tribes of Israel because the Urim and Thummim (prophetic "light-perfected" objects) were placed within the plate. The Urim and the Thummim were placed over and above Aaron's heart whenever he entered the presence of the Lord to pray and make decisions for the Israelites. For this reason, the breastplate is often called the breastplate of judgment or decision (Exodus 28:30, Romans 12, Proverbs 23:7, Psalms 24:3–4).	"DNA can turn certain genes on and off, changing the structure of neurons in the brain ... Epigenetics has revealed that our thoughts have the scientific power to change our brains. The heart's electromagnetic field is 5,000 times larger than that of the brain and the heart's light is much more significant than the brain's . . . More energy is radiated by the heart than the brain." The electromagnetic field from the heart goes out and interacts with the environment (Institute of Heartmath, Scientist).

The Stones of the Breastplate

The twelve stones were selected by God because they are multifaceted and multidimensional. They inspired many people with their beauty. But the power and symbolism of these breastplate stones go beyond simple inspiration. They were designed to store Bioglory Life-Light inside their molecules, to keep the high priest and the twelve tribes of Israel connected with the presence of God.

The book of Exodus and the Torah describe how God Himself instructed His people to make a breastplate with twelve different gemstones for the Levitical high priest to use in prayer and worship services. Like DNA, the twelve crystal stones could encode and store light energy inside their molecules. Therefore, when the ancient high priest entered the most holy place (behind the veil), the twelve crystal stones would encode and store God's intense light energy, which came from the physical manifestation of God (known as the Shekinah Glory). The molecules of the twelve stones would be filled with Shekinah Glory and resonate a rainbow-colored banner of light, and as a result, the breastplate supercharged the high priest.

In like manner, we are supercharged by the blooded priestly garment when we enter into the most holy place. Even though the naked eye cannot see the rainbow-colored banner of light that is emitted from the hexagonal-crystalline DNA, it is there. The Bible and science have proven that it is present within DNA. We can only see the rainbow-colored banner in the realm of the spirit. Many biblical writers have seen God enthroned in this rainbow banner of light:

> Immediately I was in the Spirit; and behold, a throne set in heaven, and One sat on the throne. And He who sat there was like a jasper and a sardius stone in appearance; and there was a rainbow around the throne, in appearance like an emerald.
> —Revelation 4:2-3

The appearance of the brilliant light all around Him was like that of a rainbow in a cloud on a rainy day. This was the

appearance of the likeness of the glory of the LORD. And when I saw it, I fell facedown and heard a voice speaking.
—Ezekiel 1:28

Then I saw another mighty angel coming down from heaven, wrapped in a cloud, with a rainbow above his head. His face was like the sun, and his legs like pillars of fire.
—Revelation 10:1

Heart Interference Patterns
In our bodies, much more energy is radiated by the heart than the brain, a fact that has been confirmed by the HeartMath Institute:

The heart's electromagnetic field is 5,000 times larger than that of the brain and the heart's light is much more significant than the brain's. The heart serves as a key access point through which information originating in the higher dimensional structures is coupled into the physical human system (including DNA), and that states of heart coherence generated through experiencing heartfelt positive emotions increase this coupling.

In Exodus 28, God gives instructions for Aaron on making the breastplate:

Whenever Aaron enters the Holy Place, he will bear the names of the sons of Israel over his heart on the breastpiece of decision as a continuing memorial before the LORD. Also put the Urim and the Thummim in the breastpiece, so they may be over Aaron's heart whenever he enters the presence of the LORD. Thus Aaron will always bear the means of making decisions for the Israelites over his heart before the LORD.
—Exodus 28:29-30

Knowing that the heart's electromagnetic field is 5,000 times larger than that of the brain, we can imagine that Aaron's breastplate generated powerful electromagnetic fields around his body and into the atmosphere. With the Urim (light) and the Thummim (perfection) set over his vibrating heart, his resonance was amplified. The electromagnetic field from Aaron's heart reached out and interfered with the BM-Field, radiating with the presence of God and creating overlapping patterns that rippled throughout the atmosphere. These patterns are called *interference patterns.*

To understand these interference patterns, imagine two pebbles dropped in a pond so that a ripple pattern flows from each pebble. The interference pattern will form where the two sets of waves interact and overlap to create a new pattern. If you think of these overlapping patterns as different musical notes, you could say that when we pray our heart is in harmony with the presence of God, and a perfected tune is heard in the atmosphere.

Along the same lines, I remember an old song my father used to sing: "Whisper a prayer in the morning; whisper a prayer at noon; whisper a prayer in the evening to keep your heart in tune." Prayer keeps your heart in tune with God. The heart vibrates like a musical chord when struck by His presence.

Sometimes I quiet my mind before the Lord so that my heart can resonate with His presence. I worship the Lord with every beat of my heart. I go wherever the Holy Spirit takes me. Worship and prayer begin to flow when His heart touches mine. This phenomenon is the Urim and Thummim of my blooded priestly garment in full effect.

God only cares about the heart of your prayer and worship. He looks for the Urim and Thummim on every man's heart. He inspects the heart's electromagnetic field as you worship and pray. He can tell if the electromagnetic frequency you're sending out is pure or deceitful. In Matthew 15:7-9, the Lord said:

> "You hypocrites, rightly did Isaiah prophesy of you: 'These people draw near to Me with *their mouth*, And honor Me with *their lips*, But *their heart* is far from Me. And in vain

they worship Me, Teaching *as* doctrines the commandments of men.'"

The people's hearts in this scripture are not in tune with God. There
are no *interference patterns in the atmosphere* because their hearts are far
from Him. However, we see a different story with David in 1 Samuel 16:7.

But the Lord said to Samuel, "Do not look at his appearance
or at his physical stature, because I have refused him. For
the Lord does not see as man sees; for man looks at the
outward appearance, but the Lord looks at the heart."

In addition, we are told that David, Israel's greatest king, was a
man after God's own "light-perfected" heart. God saw His very own Urim
and Thummim resting over David's heart (1 Samuel 13:14).

One day, David was out in the field with his father's sheep when a
messenger came running over to him. "The prophet Samuel is with your
family, and he wants you to come!" the messenger told him.

As David arrived at the house, Samuel anointed David as king of
Israel, much to everyone's bewilderment. Then Samuel went back to
Rameh. And where did David, the newly anointed king go? Back to the
fields to tend to his father's sheep, where he strengthened his relationship
with God through prayer and worship.

What we must learn from David's example is that spending time with
God is the most important activity on earth. "For the eyes of the Lord
move to and fro throughout the earth that He may strongly support those
whose heart is completely His." (2 Chronicles 16:9). Spending time with
God is what leads us to spiritual truth.

Intercessors with Light-Perfected Hearts

The world doesn't need more churches. What the world needs is more
men and women with light-perfected hearts, praying fervently for our
communities and nations to be touched by God. We must allow God to
alter our DNA to make our hearts light-perfected. If there were ever a

time when the world could use some people praying with light-perfected hearts, it's now.

Job asked, "What is man that You should exalt him, that You should set Your heart upon him?" (Job 7:17). God's heart is set on you. He has promised to show you great and powerful things! He has promised to come and dine with you like friends do (Revelation 3:20). God wants to share His heart with you and free you from the bondage of sin through the blood of Christ.

The blood of Christ is also known as "the Lamb's blood," which was given to the Israelites as protection. In the book of Exodus, God fights against the gods of Egypt on molecular, cellular, and quantum levels through the power of the blood:

> Now the Lord spoke to Moses (Prophet) and Aaron (Priest) in the land of Egypt. This month shall be your beginning of months; it shall be the first month of the year to you. Speak to all the congregation of Israel, saying: "On the tenth of this month every man shall take for himself a lamb, according to the house of his father, a lamb for a household…. Then the whole assembly of the congregation of Israel shall kill it at twilight. And they shall take some of the blood and put it on the two doorposts and on the lintel of the houses where they eat it…..For I will pass through the land of Egypt on that night, and will strike all the firstborn in the land of Egypt, both man and beast; and against all the gods of Egypt I will execute judgment: I am the Lord. Now the blood shall be a sign for you on the houses where you are. And when I see the blood, I will pass over you; and the plague shall not be on you to destroy you when I strike the land of Egypt."
> —Exodus 12:1-13

If you are faithful, God will pass over and above your DNA to preserve your life, just as He preserved the life of the Israelites. Conversely, when He passed over and above the Egyptians, they were destroyed—all firstborn from every Egyptian bloodline died. According to the science of epigenetics, the Lord executed modifications to the DNA within the people. These modifications preserved the life of the Israelites but brought death to the Egyptians. Epigenetics literally means to move "above" or "on top of" genetics. It refers to external modifications to DNA that turn genes on or off without altering the base genetic code. As intercessors with light-perfected hearts, when we wear the blooded priestly garment, we are also allowing God to alter our DNA and bathe us in his Bioglory.

Let's look at another example of this phenomenon: God used the Nile River as a powerful statement that the Lamb's blood was more powerful than any other power on earth. When Moses was put into a basket as an infant, he drifted along the Nile River until he reached a safe place. The literal translation of the name Moses is "one out of whom water is drawn." For the Egyptians, the Nile was a symbol of the "flow of life from the womb." They believed that their gods guarded the Nile and that their energy, power, and life flowed in the river's veins. Because of this belief, the first plague God sent came upon the Nile to dismantle the Egyptian's flow of power, overturn economic stability, and nullify life from the womb.

Thus, to unleash the first plague upon the Egyptians, Moses lifted his rod over and above the Nile River and struck it with his staff, turning the water into blood. At the same time, his brother Aaron performed an identical transformation in the canals, tributaries, ponds, and pools throughout Egypt. God started His judgment on Egypt with the first plague of blood and ended it with the last plague of blood—destroying the Egyptians at the molecular, cellular, and quantum levels through the power of the Lamb's blood.

In order to receive the blessings of the Lamb's blood, we must follow God's law, pray with light-perfected hearts, and prepare for a DNA transformation.

Joseph and the Prophecy

The pharaoh of Egypt did not know about the prophecy of Joseph or about the blooded priestly garment. Joseph's rainbow garment was the physical manifestation of a spiritual and physiological change that would occur in the DNA of the Israelites once he entered into the Egyptian's territory. And God fulfilled the prophecy through the science of epigenetics.

As told in Genesis 37:1-44, Joseph had a dream, and when he told it to his brothers, they hated him all the more and sold him into slavery. To cover up their deeds, his brothers took his rainbow garment and applied blood to it! This prophetically implied that the people of God would be freed by the blood of the Lamb. The application of the blood upon Joseph's garment activated the science of epigenetics. We previously documented that the twelve colorful stones within the high priest's breastplate represented the DNA of the twelve tribes of Israel. I believe that Joseph's rainbow garment had twelve colors, which also represented the Israelites.

The Pharaoh of Egypt was not familiar with Joseph's life story, nor was he familiar with his God (Yahweh) or the prophecy (Exodus 1:8-9).

When the Israelites were set free, they traveled out of Egypt carrying the prophecy of the blooded priestly garment over and above them. They carried the bones (DNA) of Joseph with them. God acted on their behalf through the power of the blood of the Lamb and the science of epigenetics.

> And Moses took the bones of Joseph with him, for he had placed the children of Israel under solemn oath, saying, "God will surely visit you, and you shall carry up my bones from here with you."
> —Exodus 13:19

Power in the Blood

Gatson Naessens, the renowned biologist, discovered that blood is electrically charged and contains light, energy, and frequency in the form of microscopic somatids—tiny energy particles found in the blood—which are indestructible. They are known as the smallest condensers or energy sources ever found.

Pursuing this concept, Dr. Irving added that plasma is 90 percent water and makes up more than half of total blood volume. The flow of cells is suspended within the plasma. He reasoned that blood plasma is similar to congealed light. If so, you weigh the evidence that DNA can easily encode the blooded priestly garment.

Your body is filled with tiny blood vessels called capillaries, which carry DNA-encoded messages that provide a higher color variation in humans than any other mammalian species, making your own skin a garment. We are literally covered in the Lamb's blood. All true believers are clothed in the blooded priestly garment (Revelation 1:5). This garment is not seen in the physical realm; however, it is worn by all who have been born again.

Let's take a look at another example of the power of the Lamb's blood. The blood of Christ was shed on the earth, reflecting the highest variation of colors, which fulfilled the covenant that God made with Noah and the earth. The symbol of that covenant was a rainbow.

> And God said, "This is the sign of the covenant I am making between Me and you and every living creature with you, a covenant for all generations to come: I have set My rainbow in the clouds, and it will be a sign of the covenant between Me and the earth.
> —Genesis 9:16

God never breaks His covenants. They are passed down from generation to generation. We see a similar pattern here, in the way that God has kept and fulfilled His covenants. First, we had Noah's rainbow covenant, then Joseph's coat of many colors, then Moses's blood of the Lamb, then Christ's life-giving blood that saved all mankind. Concealed within Mary's womb by the Holy Spirit, His Life-Light was encoded in Christ's DNA, creating the blood of Christ. The rainbow represents Christ's blood, and it is a sign of God's love for the earth.

Song of Solomon 2:4 says, "He brought me to the banqueting house, and his banner over me was love." The rainbow covenant shows God's love for His creation, and He will not destroy it with a flood again. Through the blood of Christ, we have been saved.

Jehovah Nissi

Christ is the banner. He is the ensign that is lifted up before the people in our communities and nations as we pray. He is Jehovah Nissi: "the Lord, my banner." When we proclaim the name of Jehovah Nissi in prayer, we are lifting up the blood-stained banner of Christ, which reflects the highest variation of colors.

Soldiers often esteem the colors of their banners for the sake of the person who first bestowed it upon them. You and I are soldiers in a spiritual battle against the enemy, and we ought to esteem the blood of Christ for the sake of God, who gave Him as a sacrifice for our sins.

David proclaims that the Lord has given a banner [token] to those who fear Him, which may be displayed because of the truth (Psalm 60:4). Again, we read in Isaiah 11:12 that "He will raise a banner [token] for the nations and gather the exiles of Israel; He will collect the scattered of Judah from the four corners of the earth."

As we pray, we are clothed with the blooded priestly garment of Christ. The power of the Holy Spirit flows through this garment, encoding the genetic language of Life-Light into our cells and transforming our DNA into the exact image of Christ. You can take this garment with you and execute your kingly and priestly duties on the earth. We are equipped by Christ to represent God in the earth realm.

In Revelation 3:18, Jesus said, "I counsel you to buy from me gold refined by fire, so that you may be rich, and white garments so that you may clothe yourself and the shame of your nakedness may not be seen, and salve to anoint your eyes, so that you may see. This is an open invitation to all mankind to commune with the Father and to lay hold of your righteous inheritance."

These white garments that Jesus counsels his followers to put on are reminiscent of the Life-Light that was emanated from His Body on the Mount of Transfiguration. While He was praying, "His face did shine as the sun," and his garments became white as light. The blooded priestly garment is a relational reality shared intimately between you and the Father as you pray. The Father is clothing many sons with the blood of the Lamb. "For the anxious longing of the creation waits eagerly for the revealing of the sons of God" (Romans 8:18-19).

This process is known as the "manifestation of the sons of God." Paul used the Greek word *apokalupsis* to describe this process, which means "the unveiling, the manifestation, or the revelation." The sons of God are manifested in the earth whenever they pray, worship, or share the knowledge of Christ (Colossians 3:10).

Concealed within the Garment

Your identity is concealed within the garment. Put it on by praying to God and unveiling your destiny. The manifestation of the sons of God on the earth is really the unveiling of sons clothed with the blooded priestly garments:

> I will greatly rejoice in the LORD, My soul shall be joyful in my God; For He has clothed me with the garments of salvation, He has covered me with the robe of righteousness, As a bridegroom decks *himself* with ornaments, And as a bride adorns *herself* with her jewels. For as the earth brings forth its bud, As the garden causes the things that are sown in it to spring forth, So the Lord GOD will cause righteousness and praise to spring forth before all the nations.
> —Isaiah 61:10

I am reminded of Esther, as she robed herself in her royal garment and went before the king to stand in the gap for her nation.

> Now it happened on the third day that Esther put on her royal robes and stood in the inner court of the king's palace, across from the king's house, while the king sat on his royal throne in the royal house, facing the entrance of the house. When the king saw Esther the queen standing in the court, she obtained favor in his sight; and the king extended to Esther the golden scepter which was in his hand. So Esther came near and touched the top of the

scepter. And the king said to her, "What do you wish, Queen Esther? What is your request? It shall be given to you—up to half the kingdom!"
—Esther 5: 1-6

Esther was born with the Hebrew name Hadassah. Her name was changed to Esther to hide her identity upon becoming queen of Persia. Esther in Hebrew means to "hide, conceal, or to be hidden." Like Esther, your identity is "hidden" within the blooded priestly garment of Christ. You have been called to the kingdom for such a time as this, to stand in the gap for your family, community, and nation.

Just like Esther was transformed from an orphan girl into a queen bride, the church was transformed from an orphan church to the bride of Christ. Through God's Bioglory, we have been genetically transformed.

CHAPTER 8
PROJECT BABEL

Come, let us build for ourselves a city and a tower whose top
will reach into heaven.
—Genesis 11:4

Mount Sinai is a cosmic mountain where the Shekinah Glory of God
rested. Moses spent time with God on this mountain, which encoded his
DNA with Bioglory, causing his face to become illuminated:

> Now it was so, when Moses came down from Mount Sinai
> (and the two tablets of the Testimony *were* in Moses' hand
> when he came down from the mountain), that Moses did
> not know that the skin of his face shone while he talked
> with Him. So when Aaron and all the children of Israel saw
> Moses, behold, the skin of his face shone, and they were
> afraid to come near him.
> —Exodus 34:29-30

In the New Testament, the Father revealed Christ's glory on Mount
Hermon, another cosmic mountain. But in the new covenant, Christ
Himself was considered the ultimate Cosmic Mountain. With his
DNA-encoded body (the temple of God), he instituted the new covenant and connected heaven to the human body. In this chapter, we will
study various kinds of cosmic mountains, and we will learn why DNA
is a gateway, the ultimate comic connection between the earth and the
heavens.

Building the Tower

In Genesis 3:15, Nimrod tries to manipulate DNA and seriously alter the bloodline of the human race during the building of the city and the tower of Babel. By using the same pattern as the crafty serpent—word and sound frequencies—Project Babel was implemented. This tower was intended to become a kind of cosmic mountain, a connection between earth and heaven.

Nimrod directed the Israelites to build ziggurat temples, which were strategically designed manmade cosmic mountains. These temples were not for worshipping God as we know Him but for worshipping the sky gods—fallen divine angelic beings who rebelled against God's divine counsel and who were cast down to the second heaven, then down to earth. Their goal was to corrupt God's creation and deceive mankind. The rituals performed in these temples were meant to maintain the cosmic order instituted by the sky gods.

The top compartment of the temple represented a "gateway for the gods" because that was where light from the sky entered into the temple. The inner walls were decorated with glazed, blue tile to reflect light from the sun, the moon, and the five zodiac planets (Mercury, Venus, Mars, Jupiter, and Saturn). The sky gods would sit on the altar in the center of the room, enthroned by light.

The tower was a place for prayer-encoding—a place where the human DNA could be illuminated by the sky god who occupied the second heaven. A person who spent time in these temples would become an "Illuminated One." Project Babel was mankind's attempt to illuminate the human genome without using God's Bioglory or Life-Light.

God considered this tower and city so repugnant that He Himself came down to get involved and put a stop to their machinations. Do you think God would have come down to earth if Nimrod and the people were only building a brick tower? No. Project Babel was detrimental to the human genome and God had to put a stop to it!

> Now the whole earth had one language and the same words. And as people migrated from the east, they found a plain in the land of Shinar and settled there. And they said to one another, "Come, let us make bricks, and burn them thoroughly." And they had brick for stone, and bitumen for

mortar. Then they said, "Come, let us build ourselves a city and a tower with its top in the heavens, and let us make a name for ourselves, lest we be dispersed over the face of the whole earth." And the Lord came down to see the city and the tower, which the children of man had built. And the Lord said, "Behold, they are one people, and they have all one language, and this is only the beginning of what they will do. And nothing that they propose to do will now be impossible for them. Come, let us go down and there confuse their language, so that they may not understand one another's speech." So the Lord dispersed them from there over the face of all the earth, and they left off building the city. Therefore its name was called Babel, because there the Lord confused the language of all the earth. And from there the Lord dispersed them over the face of all the earth.

—Genesis 11:1-9

Babel

In my opinion, human beings shared two types of languages that were confounded by the Lord at Babel: the spoken language and the DNA language.

- *The Language of Speech*: All human beings at that time shared the same spoken language, which Noah and his immediate household taught their descendants.
- *The DNA Language*: All men genetically inherited this language from Adam. The four-letter alphabet (A, G, C, and T) that makes up DNA represents a language that when transcribed and translated leads to the myriad of proteins that makes us who we are as a species and as individuals. The locations of the 98 percent of the non-coding DNA, switches, and barcodes were confounded to confuse Nimrod and the others from carrying out their plans to modify and profane the DNA of the entire human race.

God confounded both the speech language and the DNA language of the Babylonians to put a halt to Nimrod's diabolical project. The word *confound* in Hebrew translates as "to perplex, to put in a maze, to delay, to mingle, or to blend." God perplexed the map of the regulatory switches within the non-coding DNA at Babel. We previously documented that the regulatory switches control our gene expressions and mental dispositions both in the natural and spiritual realms. Therefore, it was vital for God to close off this portion of the human genome from Nimrod and the fallen sky gods.

Junk DNA or Concealed Knowledge?
Mankind lost all knowledge of regulatory switches and how they were mapped within the human genome system after Babel fell. Before the conclusions of the Human Genome Project were released in 2003, science was confused about how the human genome operated. It took over two thousand years for mankind to understand the human genome again and rise to the same knowledge level as those in ancient times.

Even at our advanced level, we are still trying to comprehend the extensiveness of the human genome system. Without knowing the correct mapping of the non-coding DNA codes and sequences, we cannot fully comprehend the human body or any other living organism. If you recite the alphabet out of sequence, you will not know the beginning or end. The right sequencing of letters A-Z is necessary for structure. In like manner, the right sequencing of non-coding DNA is necessary for gene regulation.

When Nimrod and the others tried to encode the demonic powers and genetic language of the sky gods, their efforts failed because God "mixed" and confused the regulatory switches within the human body. This was done to protect the image and likeness of God embedded in the human genome.

We can be transformed into the image of Christ and become transfigured beings of Bioglory or we can be transformed into the image of these sky gods and become transfigured beings of darkness. All of these transformations are possible because DNA has the capacity to be re-sequenced and reconstructed. God is Life-Light and Satan is the counterfeit "angel of light" (Corinthians 11:14-15).

If the wicked of Nimrod's day had gained full access to the mapping of the human genome, nothing they aspired toward would have been withheld. God protected the human genome from total distortion and genetic alterations. For this very reason, instead of going up to the skies and exploring planets for answers, humanity is now looking deep inside the cell. The cell is a universe of possibilities.

The DNA Ladder as a Cosmic Mountain

The DNA molecule is shaped like a twisted ladder, one that can literally tower into the sky like a cosmic mountain. You have about 60 trillion cells in your body, so if you placed the DNA in all of your cells in a straight line, end-to-end, they'd stretch over 744 million miles. DNA is twice the diameter of the solar system! Let's look at some more comparisons. The moon is only about 250,000 miles away, so DNA could stretch to the moon and back to earth almost 1,500 times. The sun is 93,000,000 miles away, so DNA could reach there and back about four times.

This visual—of DNA as a ladder or a cosmic mountain—can help us understand that Nimrod and the Babylonians were working on building a corrupt DNA ladder. They weren't just building a physical tower to reach God; they were trying to transform our DNA ladder as well.

A City on the Hill

In 1953, Francis Crick, James Watson, and Maurice Wilkins published a paper titled, "A Structure for Deoxyribose Nucleic Acid (DNA)." The first sentence in the article reads: "We wish to suggest a structure for the salt (SML) of Deoxyribose Nucleic Acid (DNA). DNA Double Helix is a biological salt and therefore humans are crystallized stones or precipitated light."

We previously mentioned that DNA is crystalline—a superconductor of electricity—and extremely sensitive to light waves and external energy fields. Fritz Albert Popp discovered that DNA lights up with bio-photon emissions, which display electromagnetic fields or aura-fluorescent light around the body.

I'm convinced this phenomenon was exactly what Jesus alluded to in His Sermon on the Mount:

> "You are the salt of the earth… You are the light of the world. A city built on a hill cannot be hidden… neither do men light a candle, and put it under a bushel, but on a candlestick; and it gives light unto all that are in the house."
> —Matthew 5:13-15

The Living Stone

In 1 Peter 2:4, Jesus is called the living stone. He is "the living Stone—rejected by humans but chosen by God and precious to him." The people of the world did not want this stone, but He was the stone God chose, and He was precious.

This concept is seen in the church, for we are also living stones, built up as a city of light. According to science and the scriptures, DNA is crystalline salt, which makes us the salt of the earth, and DNA is a superconductor of electricity, which makes us a city on the hill. It is clear that we are the salt, light, and stone of the earth—and as we pray, our bodies become superconductors of God's Bioglory and Life-Light.

David proclaimed:

> Lift up your heads, you gates; be lifted up, you ancient doors, that the King of glory may come in. Who is this King of glory? The Lord strong and mighty, the Lord mighty in battle. The Lord Almighty—he is the King of glory.
> —Psalm 24:7

The DNA ladder is the ancient gate of God through which His image and likeness was encoded and sequenced into Adam (Genesis 2:7). We also find that DNA is the gate through which the Word became flesh and dwelt among us (John 1:14). And lastly, DNA is the gate through which we are built up in prayer as a city on a hill that cannot be hidden (Matthew 5:14).

Genetically Modified Organisms (GMOs) or Hybrids

The DNA gates were created by God, for His divine use; however, we see that science has been modifying this gateway to profane and genetically restructure the human genome. The antichrist, a genetically modified organism (GMO) or hybrid, will use the regulatory switches within the DNA double helix to encode the power of Satan and the false prophet (Revelation 13:4).

Because of this prophecy, we can no longer perceive the building project at Babel to be a naïve, primitive effort to reach the highest heavens. Instead, this project was a scientifically brilliant but blasphemous effort to dismiss the very image and likeness of God in mankind.

Nimrod knew that if mankind prayed and opened themselves (their DNA gates) up to the sky gods, they would take on the nature and likeness of those gods because DNA can be restructured by words, sound, frequencies, and laser light beam (all epigenetic signals).

"You Will Be Like Gods"

Let's recall what Satan told Adam and Eve in Genesis 3:5: "For God doth know that in the day ye eat thereof, then your eyes shall be opened, and ye shall be as gods, knowing good and evil." This is a direct quote from the King James Version. Let's take note that the word *gods* starts with a lower-case letter "g" and the word is plural. Satan was not telling Adam and Eve that they would become like Yahweh; he made it clear that Adam and Eve would become like the fallen gods.

Once Satan and the fallen sky gods gain access to the human genome, they can regulate the genome switches and barcodes to modify human DNA, infusing it with sickness, disease, disorders, and generational curses. They can even endow human beings with demonic powers and abilities, thus reproducing a fallen image of themselves within that person.

Look around today. Don't you recognize the spiritual and human rebellion of ancient Babel? Our communities and nations are rebuilding the tower of Babel, albeit in a slightly different way.

> "Don't act corruptly and make an image for yourselves in the form of any figure . . . And don't lift up your eyes to

heaven and see the sun and the moon and the stars, all the host of heaven, and be drawn away and worship them and serve them, those which the Lord your God has allotted to all the peoples under the whole heaven. But the Lord has taken you . . . to be a people for His own possession, as today."

—Deuteronomy 4:16-20

Psalm 106 also speaks about the spiritual and human rebellion of ancient Babel: "But were mingled among the heathen, and learned their works. And they served their idols: which were a snare unto them. Yea, they sacrificed their sons and their daughters unto devils." Many served idols and sacrificed their sons and daughters to demons back then, and people are committing the same acts today. The fallen gods that were behind the building of the tower of Babel are real beings with evil intent. They led the world astray in the days of Noah and their intent has not changed. These fallen gods will do anything to keep you away from prayer and the presence of God.

I encourage you to stay connected to God in prayer. He is Light, and we are children of the Light! God has separated us from the darkness and called us His own.

God Has Allotted Evil Nations to the Fallen Gods

Moses warned ancient Israel that the nations around them worshipped sky gods as well as the hosts of second heaven. And notice his remark: God has "allotted" these nations to the "fallen gods." These nations worshipped and served the fallen gods, not the Creator God.

So when did the nations get handed over to these fallen sky gods, and why did God hand them over?

After the fall of Babel, God divided the seventy nations and placed them under the authority of the fallen sons of God. These people then became the offspring of Satan, descended from the "seed of the Serpent," and their goal was to corrupt the earth and destroy the bloodline of the Messiah.

The Most High divided the nations according to their inheritance, separating the sons of Adam and setting the DNA or bloodline boundaries of the children of Israel (His chosen people). Consider what the Lord did when He found Jacob in the desert:

> In a desert land he found him, in a barren and howling waste. He shielded him and cared for him; he guarded him as the apple of his eye, like an eagle that stirs up its nest and hovers over its young, that spreads its wings to catch them and carries them aloft. The LORD alone led him; no foreign god was with him.
> —Deuteronomy 32:9-12

Jacob's DNA Ladder

The Lord, through epigenetics, switched on genes within Jacob's DNA at Bethel (House of Prayer), fulfilling His promise to Abraham's Seed through the House of Jacob. Now let's take a deeper look into Jacob's DNA Ladder.

> And he dreamed, and behold a ladder set up on the earth, and the top of it reached to heaven: and behold the angels of God ascending and descending on it. And, behold, the Lord stood above it.
> —Genesis 28:12-13

This dream gave Jacob an internal view of his body, particularly his DNA. These angels are "messengers of light," essentially, Bioglory messengers. God is positioned at the top of the DNA ladder. The angels are ascending and descending on it in the exact way our cells send genetic messages up and down our DNA ladder to switch on or off our genetic abilities.

In the dream, Jacob grabbed hold of the Bioglory messenger, and he wrestled it, refusing to let go until his DNA was transformed. What is really happening in this wrestling match? The scriptures indicate that Jacob repented, which shows that God's BioGlory and Life-Light has the ability to change our DNA once we repent of our sins. Saul went through this same process on the road to Damascus:

> Meanwhile, Saul was still breathing out murderous threats against the Lord's disciples. He went to the high priest and asked him for letters to the synagogues in Damascus, so that if he found any there who belonged to the Way, whether men or women, he might take them as prisoners to Jerusalem. As he neared Damascus on his journey, suddenly a light from heaven flashed around him. He fell to the ground and heard a voice say to him, "Saul, Saul, why do you persecute me?"
>
> "Who are you, Lord?" Saul asked.
>
> "I am Jesus, whom you are persecuting," he replied. "Now get up and go into the city, and you will be told what you must do."
>
> The men traveling with Saul stood there speechless; they heard the sound but did not see anyone.
>
> —Acts 9:1-7

Encode: All Things New

Identity is the sum total of our distinguishing characteristics, a combination that dictates how we see ourselves. A *name* is the label this identity is given. Changing names is a small but significant step in changing identity. Jacob's name was changed to Israel and Saul's name was changed to Paul. God encoded a new name, a new nature, and a new identity into Jacob and Saul's DNA. Anyone who lives in Christ is given a new name, a new nature, and a new identity (Revelation 3:12).

Jacob used the word "Bethel" to describe God's position at the top of the DNA ladder, as His messages ascended and descended: "I have seen the face of God and lived. Confirming that God is the One to Whom our face must turn to in prayer. He alone has made all things new in our lives" (Genesis 32:31).

After meeting with God face-to-face, Jacob would later confront his fears and meet with his brother, Esau. "And Esau ran to meet him, and embraced him, and fell on his neck, and kissed him: and they wept" (Genesis 33:4).

Esau saw the Bioglory change in his brother from afar. Jacob was no longer a carnal man. He now had the language of Life-Light regulating the switches in his DNA.

You might say that I'm stretching the Word, but Jesus was even more insistent that His body was the DNA ladder, the gateway through which heaven would manifest itself on earth. In the next verse, Jesus describes His body as Bethel, an access point for Bioglory messengers (angels) to engage with those in the earth realm.

> And he saith unto him, Verily, verily, I say unto you, Hereafter ye shall see heaven open, and the angels of God ascending and descending upon the Son of man.
> —John 1:51

In like manner, your body is Bethel, an access point for heaven to invade earth with the power and glory of God. God's power and Bioglory are accessible and available to every believer. You can have a close relationship with God. The entire purpose for this book is to teach you that through prayer, you can receive and store God's Bioglory and duplicate His Life-Light on the earth.

Prayer and the Spiritual Realm

Earth's reality, everything that we perceive on the earth, comes from unseen particles. The Bible says that things that are seen are made from things that are unseen. God works with the unseen, and earth's reality can be transformed and renewed by the Word of God.

Remember when Jesus cursed the tree? It was a physical tree and Jesus spoke words to the physical tree. It dematerialized and died because Jesus spoke to the core of the tree, down to the atoms that made up the tree in its physical form. His words carried energy that affected the atoms and caused them to dematerialize, killing the tree.

The earth is continuously being transformed and renewed by the Word of God through prayer. If you pray for someone who has cancer, your words carry power. Even though the cancer has manifested in the physical realm, the Bible says that "death and life are in the power of the tongue" (Proverbs 18:21). This scripture indicates that when you pray, the energy from your words has the power to transform the spiritual and physical realms.

In Luke 8 and Matthew 8, when He and the disciples were hit by a storm while in a fishing boat, Jesus spoke to the winds and the waves. When Jesus spoke to the storm, He literally altered the reality of the earth. We can do that too because we were made in the image and likeness of God, and our words have power.

Another example of words carrying power occurs in Joshua 10. The children of Israel were in battle. Then Joshua spoke to the sun, and the sun stood still for twenty-four hours.

Our God is the king of biotechnology. Your body is His temple, a house of prayer, and heaven's gateway to earth. All you have to do is open your mouth and speak the Word of God in faith. The kingdom of God is within you. You are the New Testament temple where God Himself dwells.

Jesus described the technologies behind His New Testament covenant in John 2:19-22 when He proclaimed that His body was the ultimate temple.

> Jesus answered and said unto them, "destroy this temple, and in three days I will raise it up." Then said the Jews, "Forty and six years was this temple in building, and wilt thou rear it up in three days?" But he spake of the temple of his body. When therefore he was risen from the dead, his disciples remembered that he had said this unto them; and they believed the scripture, and the word which Jesus had said.

The same Holy Spirit that rose Christ from the dead lives within us. God is pouring out His Spirit upon the nations. As God pours out His Spirit on all flesh (DNA), we will experience a rare degree of His Bioglory beyond anything we have ever experienced. This outpouring will convince many to pursue God in prayer. Many will be drawn away from pagan worship and from the darkness and DNA alterations offered by the fallen sky gods.

The Heresy of Baal of Peor

In 1929, an archeologist excavated an ancient city lying in a large artificial mound called Ras Shamra, a city with a culture that dates back as far as 300 BC. Among the findings were literary tablets that opened the door to understanding the ancient culture of the people. Some of the tablets were Canaanite religious texts, which included detailed descriptions of the pagan deities mentioned in the Old Testament. One of those deities was Baal.

In the Bible, Baal is known as the chief sky god, and he was worshipped throughout the seven nations of Canaan. God gave His people specific instructions to tear down the high places, destroy the altars of Baal, and smash every graven image. Yet, as soon as the people settled in Canaan, they adopted the worship of Baal into their culture.

Baal is also called Horus, the son of Isis and of Osiris, known as "the one far above" or "the sky god" of the cosmic mountain in ancient Egyptian worship. In modern culture, we see Horus being worshipped as the one who illuminates DNA. Horus and many of the other sky gods have human forms that have been mixed with animal and angelic genes to create GMOs.

The worship of Baal was so prevalent throughout ancient Israel that God had to intervene with dramatic displays of His authority and power in the earth realm. In those days, Elijah had to face Ahab, the most wicked king who ever reigned over Israel. Ahab married Jezebel, who worshiped Baal. Jezebel started to kill the prophets of God, and both she and Ahab created problems for the prophet Elijah. Not only did Elijah have to fight against the enemies of God, but he also had to convince the Israelites to stop integrating Baal into their daily worship routines. The people of God had a powerful attraction to Baal. Elijah had to deal with paganism, witchcraft, and sorcery in all levels of Israelite society:

And Elijah came to all the people, and said, "How long will you falter between two opinions? If the LORD is God, follow Him; but if Baal, follow him." But the people answered him not a word. Then Elijah said to the people, "I am the only remaining prophet of the LORD, but Baal has four hundred and fifty prophets."

—1 Kings 18: 21-22

Jezebel's Occult Prayers

Jezebel was a Phoenician princess, which indicates that she was well educated in the culture of Baal and occult worship. She was the only queen named in the list of Israel's kings because her dark powers undergirded King Ahab's throne. He made no decisions without first consulting Jezebel.

Have you ever wondered why Jezebel wanted to kill all of God's prophets? Because prophets represent the voice of God in the earth realm. When prophets speak, they speak the sound and frequency of heaven, and Jezebel was a follower of Baal. She wanted to control the sound frequencies for Baal. Her name literally means "to summon Baal" with a loud, ritual cry!

Jezebel mastered the art of occult prayers in the high places of the earth. The high places are the seven pillars of society: the church/religion, government, business and economy, education, science and technology, sports, and arts and culture.

Today, the church is called to take back and occupy the high places of the earth. Deuteronomy 33:29 says, "Happy are you, O Israel! Who is like you, a people saved by the LORD, The shield of your help And the sword of your majesty! Your enemies shall submit to you, And you shall tread down their high places." The high places are always the most desirable places in society to occupy, and they are the best places to erect physical and spiritual altars to commune with the gods (1 Kings 14:23). From these altars, Jezebel and her false prophets would send occult frequencies over the nation.

Remember that sound frequencies can reprogram DNA because DNA is always listening for instructions to encode, store, and reproduce. The cry of Jezebel went out, over and above the DNA of the people, to modify their psychological processes and ultimately sway the minds of the people. Jezebel's frequency is a binaural ritualistic sound that brings confusion and division within heads of governments, communities, and nations. This is why Elijah asked the people: "How long will you be confused as to who is your God?" (1 King 18:21).

Elijah had to stand in the midst of a generational occult voice and fight against it with the power of prayer. He was not just fighting Jezebel and Ahab. He was fighting against the very gates of hell. His assignment was to restore the altar of God in the high places. Mount Carmel is a cosmic mountain, a location where heaven connects with earth. But Jezebel had closed up this gateway of prayer and worship unto Yahweh by tearing down the altar of God. She then built altars to Baal and Ashtaroth, giving them access to the people and nation.

Today, Jezebel, also known as the "great whore of Babylon," is still doing Baal's work. She is still opening her mouth to send out frequencies that go over and above DNA and modify the psychology of the masses. In Revelation 17:1-18, John pictures the whore of Babylon sitting on seven hills (seven high places or pillars in society), and this scripture shows that Project Babel and the manipulation of the human genome is still at work in the end times:

And he said to me, "The waters that you saw, where the whore is seated, are peoples and multitudes and nations and languages. And the ten horns that you saw, they and the beast will hate the whore; they will make her desolate and naked; they will devour her flesh and burn her up with fire. For God has put it into their hearts to carry out his purpose by agreeing to give their kingdom to the beast, until the words of God will be fulfilled. The woman you saw is the great city that rules over the kings of the earth."

We must fight against these occult voices in the realm of the spirit by "casting down arguments and every high thing that exalts itself against the knowledge of God, bringing every thought into captivity to the obedience of Christ" (2 Corinthians 10:5). We are called to dispossess not only our minds but our nations of these occult voices (Ephesians 3:10-11).

This is what Elijah was summoned by God to accomplish. He was called to dispossess the land of Baal and the occult voice of Jezebel, so the Lord could repossess the mind and hearts of the people.

Tearing Down Pagan Altars

Elijah was instructed by the Lord to occupy the high places and tear down the altars and images associated with Baal and Ashtaroth. In like manner, every leader or shepherd in the kingdom of God is commanded to do the same. These leaders will be individually evaluated as to whether they have obeyed God's command to occupy high places and cut down the pagan altars in our society. God's commands have not changed. His Word is forever settled in heaven (Psalms 119:89). He has instructed His prophets, priests, and kings on earth to tear down the high places and repair His prayer altars.

For example, in 2 Chronicles 14:2-5, King Asa did what was right in the eyes of the Lord, for he removed the altars of the foreign gods, broke the sacred pillars, and cut down the graven wooden images in all the high places. He commanded the people of Judah to pray and seek the Lord God of their fathers and observe His laws and commandments. He also removed the incense altars from all the cities of Judah, and the kingdom was quiet under him.

In addition, his son Jehoshaphat walked in all the ways of his father. He did not turn aside from doing what was right in the eyes of the Lord. Nevertheless, he did not instruct the people to tear down the false altars in the high places, and the people offered sacrifices and burned incense in those high places (1 Kings 22:24-43). Later, another king, King Hezekiah, received an even more favorable evaluation. Because King Hezekiah actually removed the foreign altars, he found favor in the sight of the Lord (2 kings 18:2-4).

Let's look at one more example. In 2 Chronicles 34, the Israelites were not following the command of God to tear down the high places. Then *Hilkiah* found a lost copy of the Book of the Law and read the law to King Josiah. When Josiah heard the Word of God, he tore his clothes in a common expression of grief and anguish (2 Chronicles 34:19). According to the scriptures, his father, Amon, was a sinful king and an idolater. But King Josiah broke down the altars of the Baal in his presence: the incense altars, the wooden images, the carved images, and the molded images. He turned them into dust and scattered them on the graves of those who had been sacrificed in Baal's name. He also burned the bones of Baal's priests and thoroughly cleansed Judah and Jerusalem.

The point is that the kings who tore down the high places were the kings most commended by God.

Repairing the Prayer Altar

Jezebel and her false prophets tore down the altar of the Lord in the high places, and they are still working on tearing down that altar today—but on a spiritual level. I posit that the spiritual altar of the Lord is your DNA ladder. Let's study this spiritual truth.

In ancient history, the altars of God were never made with stones that were crafted by human hands:

> And if you make an altar of stone for Me, you shall not build it of cut stones, for if you wield your tool on it, you will profane it.
> —Exodus 20:24-25

DNA is also something that is not made by human hands. The New Testament altar of the Lord is your DNA ladder, which represents the following:

- A place of sacrifice and worship – People go to the altar to offer sacrifices and worship God (Genesis 22:9; Ezekiel 6:3; 2 Kings 23:12; 16:4; 23:8).

- A place that reminds us of God's covenants and promises – Abraham built an altar to remind him of the covenants and promises that God made unto his bloodline (Genesis 13:4; 22:9).
- A place of purification and prayer – At the altar, many people of God have used the power of prayer to purify the people (Isaiah 6:6; Acts 8:1-24).
- A place to receive gifts from God – Through the Holy Spirit, we can receive a constant stream of energy that emanates from the throne of God (Daniel 7: 9-10; 1 Thessalonians 5: 19; 2 Chronicles 7:15; Matthew 18:20).
- A place where heaven is connected to the earth – Both physical and spiritual altars are gateways to heaven (1 Kings 18: 30).

In other words, your body is a temple and an altar to God—a place of worship and meditation where you can spend time in His presence. At this spiritual altar, you can also renew your covenants with Him. Because of these covenants, you have the right to speak against things that are not in line with God's commandments. Your lips are purified, just like Philip's lips were purified in Acts 8:1-24. When Philip entered Samaria, the demons ran before him, expelling themselves from possessed bodies because they were terrified of the Life-Light within him. Philip spoke the Word of God, and he was able to convert many people. You will also be able to receive God's gifts, including the famed "tongues of fire" that many prophets have used to transform people, communities, and nations. Lastly, your DNA connects heaven to earth, allowing the Lord direct access to your body to bestow generational blessings, power, and glory upon you.

Romans 12:1, says, "Therefore, I urge you, brothers and sisters, in view of God's mercy, to offer your bodies as a living sacrifice, holy and pleasing to God—this is your true and proper worship. Do not conform to the pattern of this world, but be transformed by the renewing of your mind. Then you will be able to test and approve what God's will is—His good, pleasing, and perfect will."

As 1 Peter 2:5 proclaims, "You also, like living stones, are being built into a spiritual house to be a holy priesthood, offering spiritual sacrifices acceptable to God through Jesus Christ."

The Fire of God

In order to release the "river of fire" and the "tongues of fire" from His throne, God needed the altars on earth repaired. After Elijah restored His altars, the Lord poured out His Glory upon the people. "Then the fire of the LORD fell and consumed the burnt sacrifice, and the wood and the stones and the dust, and it licked up the water that *was* in the trench" (1 Kings 18:38). This fire was not lightning; it was a supernatural Life-Light emanating from God Himself, which saturated the altar and purified it.

God has promised to do the same when we repair His altars of prayer and worship in our lives, communities, and nations. The word *repair* means "to reestablish something to its original condition." Therefore, when you have been born again, your DNA has been repaired to its original condition. Your spirit, soul, and body will function according to God's original design by encoding, storing, and duplicating His Bioglory.

When something has been repaired or restored in scriptures, it always grows, multiples, or improves in some way so that its final condition is superior to its original state (Joel 2:21-26, Psalm 51:12). For example, when God restored the human genome, He replaced the old Adam with a more superior second Adam—that is, Christ—a life-giving Spirit (1 Corinthians 15:45). Again, when God restored Job, he gave him double that which he had lost and blessed him more abundantly in his final years than at the beginning of his life (Job 42: 10-12).

God multiplies when He restores, and thus, in modern times, God has not only blessed us with tongues of fire, but He has also promised a greater outpouring of the Holy Spirit (Acts 2:17).

Before we venture into the next chapter, I leave you with these encouraging words: Even though Project Babel is still in operation today, Christ in the flesh has completed the work of His Father. Part of His work was to restore the human genome back to its original condition. He accomplished that goal and so much more. If the rulers of the world understood the mysteries of the human genome, they would not have crucified the Lord of Glory. Like energy, the Life-Light within Christ cannot be destroyed, but it can be transferred to his followers.

Today, the Holy Spirit is taking His work within our DNA to the next level of transformation—our final condition will be far superior to anything Project Babel could have produced. In Christ, we have been rescued, redeemed, restored, and forgiven (Colossians 1:13-14).

CHAPTER 9
DNA IS A PHOTOCOPIER

And do not be conformed to this world, but be transformed by the renewing of your mind, that you may prove
what *is* that good and acceptable and perfect will of God.
—Romans 12:2

Imagination is your ability to capture your thoughts in picture form. God
has built imagination into your genetic makeup. When you "cast down
vain imaginations," you are literally dispossessing your mind of any and
all evil "mental imagery" that is trying to reshape the blueprint of your
genetic makeup.

We must capturing and pull down vain imaginations so we can calm
down our spirits and tune into the voice of God. "You will keep him in
perfect peace, Whose mind is stayed on You, Because he trusts in You"
(Isaiah 26:3).

Whatever is true, whatever is noble, whatever is right, whatever is
pure, whatever is lovely, whatever is admirable—think about such things
because you will become like the images in your mind. Everything in
the realm of the spirit needs a pathway to access the earthly realm. You
become what you think. So why keep capturing vain imaginations?

In this chapter, we will discuss the vain imaginations that keep you
from God. These thoughts and ideas are not gifts from God, but rather,
curses from Satan.

DNA Copies Itself

Growing up in the Bahamas, we did not watch very much TV. We kept
ourselves entertained with our imaginations. We often pretended to be
famous people, and I would always choose Whitney Houston. I would
go into my closet and put on my best dress and shoes to grace our make-
believe stage. I would pretend that I was signing in front of a sold-out

crowd. At that time, no one could have told me that I wasn't Whitney Houston. I could mimic her stage performance—every move, every note, every word—because I saw her in my mind.

Our spirits do the same thing. Whatever images we hold in our minds, our brains will mimic and copy within our neurons. Thus, images can create substance. In addition, DNA is a photocopier. It makes a copy of the coded images, which then serves as a guide or an architectural blueprint for our bodies to follow. Thus, we begin to act, speak, and think like that image. You will always be transformed into the image set within your mind. This phenomenon is called *genetic expression* or *spiritual expression*.

The good news is that we can capture the mind of Christ in picture form, and we can focus on it through meditation. By focusing on His glory, we can become like him. However, if we are not focusing on the mind of Christ, then our minds can be snared by the enemy (2 Timothy 2:26). A *snare* is a trap or a strategy set by the enemy to capture you and keep you captive. Paul said that we often wrestle against spiritual wickedness in heavenly places. Snares for your thoughts can be set when these wicked spiritual beings try to counsel your mind by overriding your mental disposition. Do not allow Satan and his followers to counsel your mind and interject their thoughts. These are vain imaginations.

Vain imaginations start with a fleeting thought, but if you do not take that thought captive immediately, it can form a mental picture in your mind. That mental picture will then flow through your neurons and through the pathways of your mind, cementing itself within them. You will begin "seeing" your lusts, fears, anxieties, and worries play over and over in your mind. Vain imaginations elicit depression, discouragement, stress, anxiousness, worry, and doubt—all of which originate from the enemy. The enemy is a master of deception and can present you with thoughts and feelings that seem so real, but they aren't. You can make them real by accepting and processing them through your neurological pathways.

Because your DNA is sensitive, all the things in your mind—from the thoughts you think to the words and sounds you hear—are co-creating your life. Your thoughts determine the electromagnetic signals that reach your brain.

For example, if you're having a bad day and you find yourself in a negative frame of mind, this electromagnetic signal will generate a low frequency impulse throughout your body, and your non-coding DNA will respond to this impulse by "switching off" certain hormonal pathways in your brain. As a result, you will feel downcast, frustrated, and depressed.

On the other hand, if you're having a bad day, and you are able to pray the Word of God and break out of your negative mindset, a high frequency electrical signal will reach your DNA, and you will feel lighter and more joyful. You will feel like a heavy burden has been lifted off of your shoulders. Your DNA will respond to the Life-Light in God's Word by "switching on" hormonal signals in your brain that will make you feel better.

Remember, even a slight shift in your thought processes can trigger a response from your DNA. So, try to capture any negative thoughts right away and pray your way to a better day.

Spiritual Warfare is Holistic in Nature
You can fight battles in your mind by actively programming and encoding your own DNA with the Bioglory of God. The core of our spiritual battles are fought through the science of epigenetics. The environment of your mind is the battlefield, and this environment affects your genes.

Spiritual warfare is far more holistic in nature than what you may have been taught. The mind's environmental impact extends to the electromagnetic world of quantum physics, your non-coding DNA, and the switches that regulate your gene expressions. All of these unseen elements combine to reveal your reality.

Spiritual warfare begins in the mind. However, the mind is intimately connected with the body, so the results of the battle will affect the whole person, taking into account mental, physical, and social factors.

In the following sections, we will learn more about spiritual warfare and how through prayer, you can create the optimal mental environment to unlock the highest potential in your DNA. This is the great secret that praying with the language of glory holds: spending time with God in His Secret Place can alter your DNA and prepare you to do His will on earth.

Let me show you how to abide under the shadow of His Almighty Grace, Love, and Mercy.

The Power of Praying with the Language of Glory

For a period of time in my life, I got lost in a negative frame of mind, and the electromagnetic signal from my thoughts generated a low frequency impulse throughout my body. As a result, I became very depressed and suicidal. Spending time in the presence of God changed my life and my mental environment. His presence became my refuge and a place of peace. Today, I am no longer suffering from depression or the suicidal thoughts that the enemy injected into my mind. God's Bioglory has improved my metal disposition and encoded my DNA. My place of prayer has become my place of power and transfiguration.

My dear reader, if you are suffering from depression or suicidal thoughts, I urge you to discover the presence of God for yourself; your life will be transformed before your very eyes. In His presence, you will experience "good vibrations." The Holy Spirit is on standby to provide you with good counsel. He is your advocate, comforter, intercessor, strengthener, and friend.

I can truly identify with what you're going through. Sometimes we feel like the weight of the world is on our shoulders. We might feel like no one else really understands what's going on in our minds. We often feel alone and deserted. But I can tell you that the Holy Spirit understands. He sees your thoughts and He feels your inner hurt. He can hear your faintest cry. He is right there beside you. Allow Him to be your counselor and friend. He will lead you to the path of peace, joy, and wholeness.

Be still. Be at rest in your mind. If you can't find the words to pray, the Holy Spirit will pray for you. Romans 8:26 says that the "Spirit Himself intercedes for us with groans too deep for words." The Spirit of God is the same yesterday, today, and forevermore. With God, "there is no variation or shifting shadow" (James 1:17). In other words, you can rely on Him. Though your experience changes from day to day, hour to hour, minute to minute, the Holy Spirit will never change, and His counsel to you will always remain the same. He is your ever-present helper. He will never leave you, nor will He forsake you. He will not fail you. Trust His counsel, and you will have victory in this battle.

Dr. Cindy Trimm is a life strategist, humanitarian, and best-selling author. In her book, *Rules of Engagement: The Art of Strategic Prayer and Spiritual Warfare*, she explains that "spiritual warfare is the counsel of the human mind by any other spirit (including the human spirit) other than the Spirit of God."

Dr. Trimm then wrote about Matthew 4 to give insight on how the enemy came to Jesus in the wilderness to attack His mind. Jesus was led up by the Spirit into the wilderness to be tempted by the devil. The voice of the enemy spoke to Him on three occasions:

> After He had fasted forty days and forty nights, He became hungry. And the tempter came and said to Him, "If You are the Son of God, command that these stones become bread." But He answered and said, "It is written, 'Man shall not live on bread alone, but on every word that proceeds out of the mouth of God.'"
>
> Then the devil took Him into the holy city; and he had Him stand on the pinnacle of the temple, and said to Him, "If You are the Son of God throw Yourself down; for it is written, 'He will give His angels charge concerning You'; and 'On their hands they will bear You up, lest You strike Your foot against a stone.'"
>
> Jesus said to him, "On the other hand, it is written, 'You shall not put the Lord your God to the test.'"
>
> Again, the devil took Him to a very high mountain, and showed Him all the kingdoms of the world, and their glory; and he said to Him, "All these things will I give You, if You fall down and worship me."
>
> Then Jesus said to him, "Begone, Satan! For it is written, 'You shall worship the Lord your God, and serve Him only.'" Then the devil left Him; and behold, angels came and began to minister to Him.

Jesus used the Word and brought His thoughts under subjection to the authority of His Father. You must do the same. Once the devil takes your mind, he has authority and dominion over you. You must fight every thought and every high thing that does not align with the Word of God and the authority of Christ.

Whenever we fail to cast down vain imaginations, we begin to suffer from fear, worry, and all manner of unstable emotions that will send out the wrong signals throughout our bodies, right down to the molecular level of our DNA. God has designed the human body with the ability to accomplish His will. He wants us to observe our thoughts, catch those that are bad, and cast them out of our minds so that we may be comfortable and joyous in His presence.

The importance of capturing God's Word and casting down vain imaginations cannot be underestimated, as Proverbs tells us:

> My son, pay attention to what I say; turn your ear to my words. Do not let them out of your sight, keep them within your heart; for they are life to those who find them and health to one's whole body.
> —Proverbs 4:20-22

You Will Become Like the Image in Your Mind

Research shows that the vast majority of mental and physical illnesses come from the unstable imaginative processes of the mind rather than from the environment or hereditary genes. James 1:8 explains, "A person who has a double mind is thinking about two different things at the same time and can't make up his or her mind about anything." Meaning, your vain imaginations can be in conflict with reality. As we explained earlier in this chapter, Satan's lies will fight against the truth of God's Word in your life.

We are wired to dispossess our mind of all evil and vain imaginations. Research has shown that five to ten minutes a day of intentional, meditative capturing and casting down of vain imaginations can shift the frontal brain states so that we are more likely to engage with our surroundings and have a healthier mental outlook on life.

Interceding for the Long Haul

Elijah was completely in tune with God's plans for Israel. He listened for the voice of God, and he walked in obedience (1 Kings 18:36). He prayed in a way that pleased God, and his prayers pointed the nation back to Him. He prayed, "O Lord, answer me! Answer me so these people will know that you, O Lord, are God and that you have brought them back to yourself" (1 Kings 18:37).

Elijah prayed fervently until the powers of darkness were broken. He knew that he had to stand in prayer until the river of fire came down from the throne of God. Elijah prayed until he saw the supernatural break through the veil. And he was committed to intercession for the long haul. The spiritual battle and physical display of fire shifted the minds of the people away from Baal and back to God.

> The people shouted with one voice, When all the people saw this, they fell facedown and said, "The LORD, He is God! The LORD, He is God!"
> —1 Kings 18:39

I bring up the story of Elijah again because it shows how vain imaginations can be in direct conflict with the truth of God's Word. The minds of the people were clouded by vain imaginations; Elijah, through prayer, was able to refocus the minds of the people on the one true God.

As we explored in chapter six, neurons are nerve cells with axons and dendrites that carry electrical signals throughout the field of the brain. These neuroelectrical pathways in the brain look and perform like trees— their branches bearing fruit when aligned with the mind of Christ. If we fail to think good thoughts pleasing to the Lord, our neuroelectrical pathways become a conduit for the enemy's lies. When we allow our neuroelectrical pathways to carry fears, discouragements, doubts, and worries, our trees become distorted because the energy within these neurons releases harmful epigenetic signals in the brain that shape the reality of our lives, our mental dispositions, and our physical health. In other words, the state of your mind directly affects the state of your brain, and the state of your brain directly affects your physical life.

Make God the object of your faith and worship, and soon, you will reflect His Bioglory and Life-Light. When you pray with the language of glory, paradigms will be shifted and destinies changed.

The Object of Your Faith

Having faith is important. But more important than having faith is choosing the right *object of your faith*. There is no doubt that the four hundred prophets of Baal and four hundred fifty prophets of Asther had real faith. They were cutting themselves and earnestly crying out to Baal for hours. "But the was no voice; no one answered" (1 King 18:26).

There was nothing wrong with their faith, but the object of their faith was unworthy. They had faith in Baal; he was their god. But their faith was worthless because the object of their faith was a false god. The great things that they hoped would manifest on Mount Carmel never came to pass.

Baal is a fallen angel, which means he has substance. But what type of substance can a false god have? Baal's substance is evil, and he cannot stand up against the righteousness of Yahweh. He cannot display any power that is above the abilities of Yahweh. Baal's power is void of substance when it comes to the power and authority of God.

The proof is in the substance. The best way to find out if something has a true and good substance is to test it yourself. In Kings 18, Elijah called Ahab to witness the power of God for himself.

> Elijah climbed to the top of Mount Carmel and bowed low to the ground and prayed with his face between his knees. Then he said to his servant, "Go and look out toward the sea . . .
> Finally the seventh time, his servant told him, "I saw a little cloud about the size of a man's hand rising from the sea."
> Then Elijah shouted, "Hurry to Ahab and tell him, 'Climb into your chariot and go back home. If you don't hurry, the rain will stop you!'"

And soon the sky was black with clouds. A heavy wind brought a terrific rainstorm, and Ahab left quickly for Jezreel.
—Kings 18:42-45

Ahab witnessed the power of God. The God of Elijah accomplished a miracle that Baal was unable to perform. By the power of God, Elijah had prayed the drought into existence, and with that same power, he prayed the drought out of existence. He oriented his mind to accept the will of God. Throughout the spiritual battle on Mount Carmel, Elijah kept his mind focused on God, and he was rewarded.

When we orient our minds toward God's mind, we can demonstrate His power on the earth because our DNA is a photocopier. We are wired to display God's power and glory through prayer. During prayer, our DNA encodes the thoughts and Life-Light of God, which rewires our brains and transforms our lives so that we too can display the kind of power and authority demonstrated by Elijah.

Elijah made a clear distinction between Yaweh, "*the* Cloud Rider," and Baal who is a worthless imposter, "*a* cloud rider" (Psalm 104:3-4). Elijah made it clear that "Baal is nothing—and nothing can come from him."

As we pray, we must remember that it is God's faithfulness that we are calling upon, His character. God takes pleasure in the display of His power and glory before the wicked. He delights when we stand with Him in faith.

The Lord said, "Because he loves Me, I will deliver him; because he knows My name, I will protect him. When he calls out to Me, I will answer him; I will be with him in trouble. I will deliver him and honor him."
—Psalm 91:16

Dispossessing the Land of Baal

When Elijah repaired the altar of the Lord by calling down the river of fire against Jezebel, Baal, and the false prophets, it was a sure sign that he was appointed by God to speak and demonstrate His power. Whenever God's power is demonstrated on earth, evil is physically and spiritually dispossessed from the land.

Through Elijah, God not only physically dispossessed the land of false prophets, but He also dispossessed the high places of Baal. Elijah overthrew Baal's culture, as opposed to blending in with it, and every prophet after Elijah stood in opposition to Baal's culture and worship. Hosea described the adulterous intimacy that both Judah and Israel had with Baal, and Jeremiah battled with an infestation of it in Judah (Hosea 2:13, 16-17 and Jeremiah 2:23; 32:35). Therefore, Elijah had to battle with ingrained images that were imprinted into the people's DNA. Those vain imaginations had been carrying on in that area of the world for generations.

This era is a crucial time for believers to stand, watch, and pray—just like the sons of Issachar prayed during their era. These men were from the renowned tribe of Israel and served as counselors to King David. They understood that their time and season required staunch opposition to Baal (1 Chronicles 12:32). In the same way, we are to pray during these dark times and seasons, so we can overtake evil men and their worldly wisdom.

Peter said that God has given us precious and magnificent promises. Through those covenants, we can become partakers of His divine nature (2 Peter 1:4). With everything going on in these times, we must not forget that we have made a covenant with Christ. We are partakers of His divine nature, and we should never allow our DNA to be profaned by the "idolatrous wine" of Babylon. We must come out of Babylon, for we are in this world but not of its sinful culture. We must remain in prayer, connected to God.

> For all the nations have drunk the maddening wine of her adulteries. The kings of the earth committed adultery with her, and the merchants of the earth grew rich from her excessive luxuries. Then I heard another voice from heaven say: Come out of her, my people, so that you will not share in her sins, so that you will not receive any of her plagues;

for her sins are piled up to heaven, and God has remembered her crimes.

—Revelation 18: 3-5

In this scripture, an angel is addressing the church of God, even as Moses addressed the Israelites, and he is warning them not to engage with the sky gods of the neighboring Canaanite nations.

You can be assured that God remembers the crimes of ancient Babel, and he witnesses the crimes of modern Babylon. There is a DNA boundary that Satan and the fallen sky gods continue to try to lead men across. Satan's goal remains the same: to profane and intoxicate the temple of God. Your body is God's new covenant temple, and you must work to keep it pure. We are called to be kings and priests, and we must submit to God's will. Any time we fail to submit to God in prayer, we profane God's temple by opening a doorway to compromise. For that reason, we must be vigilant and steer away from any vain imaginations that could replicate negativity within us. We must align our minds with the mind of Christ.

CHAPTER 10
CROSSING DNA BOUNDARIES

God stands in the congregation of the mighty; He judges among the gods . . . I said, "You *are* gods, And all of you *are* children of the Most High. But you shall die like men, And fall like one of the princes."
Arise, O God, judge the earth; For You shall inherit all nations.
— Psalm 82:1, 6-8.

During and after the great flood, fallen angels were having sexual relations with human women who gave birth to children. These offspring were considered hybrid—half-human and half-angel—and they were known as Nephilim. These children became the "men of old" or "men of renown" and the legendary figures of ancient times.

Now it came to pass, when men began to multiply on the face of the earth, and daughters were born to them, that the sons of God saw the daughters of men, that they *were* beautiful; and they took wives for themselves of all whom they chose. And the LORD said, "My Spirit shall not strive with man forever, for he *is* indeed flesh; yet his days shall be one hundred and twenty years." There were giants on the earth in those days, and also afterward, when the sons of God came in to the daughters of men and they bore *children* to them. Those *were* the mighty men who *were* of old, men of renown.
—Genesis 6:1-4

The origin of the giant offspring is of fundamental importance to our study of prayer. This passage presents a startling message that actual fallen sons of God (angels) mated with humans to create terrible hybrids as tall as giants. It was the ultimate corruption of the gene pool. This revelation helps us in our prayer lives because we can now understand the judgment Christ brought upon the Nephilim, and we can see how he expects us to uphold His victory in the end times.

When we pray, the phrase "standing in His victory" is often quoted, but do we truly understand what it means to stand in the victory of Christ over these Nephilim? Standing in His victory means that we are fully identified in the knowledge of His creation. We know that through His sacrifice, He restored the language of Bioglory to the our DNA. With this knowledge, we pray, share the Gospel, and occupy the high places of the earth (Ephesians 3:10).

In this chapter, we will explore how Satan and his followers crossed DNA boundaries and continue to cross DNA boundaries by corrupting the gene pool. We will also learn how we (as followers of Christ) can uphold God's victory and dominion over the earth.

Illegitimate Offspring

Because God gave dominion of the earth unto man, the Nephilim were illegitimate "seeds" in the earth realm. The human body was made by God and consecrated unto Him alone. The fallen sons of God crossed DNA boundaries when they slept with human women and produced offspring. This act was a gross violation of God's consecrated temple (Genesis 1: 26-28; Deuteronomy 23:1; Leviticus 27:28; Numbers 18:14; Joshua 6:18; Micah 4:13).

Satan knew that the Messiah would use human DNA as a gateway to redeem and restore the human genome and encode God's Bioglory. But he also knew that if human DNA was mixed with angelic DNA, the bloodline leading to the birth of the Messiah could be corrupted, leaving the human race no chance for salvation.

Why would a loving Creator destroy humanity with a cataclysmic global flood? Because too many DNA boundaries had been crossed. Our loving God made man in His very own image and likeness, and he wanted man to have dominion over the earth. The fallen sons of God corrupted the human gene pool, and the only way to save the human race was to remove the corrupted bloodline.

The Book of Jude also documents the crossing of DNA boundaries. "Sodom and Gomorrah, and the cities around them in a similar manner to these, having given themselves over to gross immorality and gone after strange flesh, are set forth as an example, suffering the vengeance of eternal fire"(Jude 1:7).

We often correlate the "gross immorality" mentioned in this scripture with homosexuality, but the offenses of Sodom and Gomorrah go much deeper. In those cities, the sin was not that men were going after men or that women were going after other women. Rather, the sin was that they were pursuing angels:

> That evening the two angels came to the entrance of the city of Sodom. Lot was sitting there, and when he saw them, he stood up to meet them. Then he welcomed them and bowed with his face to the ground. "My lords," he said, "come to my home to wash your feet, and be my guests for the night. You may then get up early in the morning and be on your way again."

> "Oh no," they replied. "We'll just spend the night out here in the city square."
> But Lot insisted, so at last they went home with him. Lot prepared a feast for them, complete with fresh bread made without yeast, and they ate. But before they retired for the night, all the men of Sodom, young and old, came from all over the city and surrounded the house. They shouted to Lot, "Where are the men who came to spend the night with you? Bring them out to us so we can have sex with them!"

> So Lot stepped outside to talk to them, shutting the door behind him. "Please, my brothers," he begged, "don't do such a wicked thing. Look, I have two virgin daughters. Let me bring them out to you, and you can do with them as you wish. But please, leave these men alone, for they are my guests and are under my protection."

"Stand back!" they shouted. "This fellow came to town as an outsider, and now he's acting like our judge! We'll treat you far worse than those other men!" And they lunged toward Lot to break down the door.

But the two angels reached out, pulled Lot into the house, and bolted the door. Then they blinded all the men, young and old, who were at the door of the house, so they gave up trying to get inside.
—Genesis 19: 1-11

The people of Sodom and Gomorrah were sinning in the same way the people were sinning in Genesis 6; they were crossing DNA boundaries. As a result, God sent the flood. Today, many people are pursuing the same kind of immorality, which will bring down the fire of God upon us.

The Days of Noah
When discussing the time near His return to earth, Jesus declared, "But as the days of Noah were, so also will the coming of the Son of Man be" (Matthew 24:37). During the days of Noah, uncorrupted human DNA was rare. Outside of Noah's immediate family, every other human bloodline had been profaned by the fallen sons of God.

The Bible reveals that in the end times, prior to Christ's return, people will also ignore warnings of the same message, rejecting the Gospel of Christ and accepting the doctrines of devils. The doctrines of devils are founded on the primeval sin of Genesis 6. As we pray, we must take up the sword of the Spirit (Word of God) and take our stand against the fallen angels and the hosts of darkness. While we prepare to fight in the realm of the spirit, the Lord will give us strategies on how to stand against the enemy.

Ancient Prayer Strategies
The ancient Israelites were ready to wage war against the giants and the Canaanites, who controlled access to the Dead Sea, the Jordan River, and

the Sea of Galilee. But the Lord would not allow them to fight without first giving them a strategy. The Lord instructed the Israelites to first bind the gatekeepers (the Nephilim kings):

> And the LORD said to me, 'See, I have begun to give Sihon and his land over to you. Begin to possess *it*, that you may inherit his land.' Then Sihon and all his people came out against us to fight at Jahaz. And the LORD our God delivered him over to us; so we defeated him, his sons, and all his people. We took all his cities at that time, and we utterly destroyed the men, women, and little ones of every city; we left none remaining. We took only the livestock as plunder for ourselves, with the spoil of the cities which we took. From Aroer, which *is* on the bank of the River Arnon, and *from* the city that *is* in the ravine, as far as Gilead, there was not one city too strong for us; the LORD our God delivered all to us.
> —Deuteronomy 2:31-36

Every step of the way, God went out in front of the Israelites' army to fight the initial spiritual battle and severely cripple the enemy so His people could finish the battle in the physical realm. God destroyed Sihon and his sons—the entire bloodline of wicked gatekeepers.

Binding the Gate Keepers

Wicked gatekeepers are physical and spiritual barriers. In order for you to possess what's rightfully yours, you must overcome them. I pray that God will overturn every wicked gatekeeper for your sake, giving you access to your inheritance. As previously discussed, God will specifically and strategically destroy any and all demonic forces that have risen up against His people. So, we must tune our ears to His Voice and follow His prayer strategies.

God said, "But as my people watched, I destroyed the Amorites, though they were as tall as cedars and as strong as oaks. I destroyed the fruit on their branches and dug out their roots. It was I who rescued you from Egypt and led you through the desert for forty years, so you could possess the land of the Amorites" (Amos 2: 9-10).

After defeating Sihon, the Israelites faced the second gatekeeper, a Rephaim giant, King Og of Bashan. King Og is one of the most famous of the post-Flood gatekeepers. His regions covered more than sixty cites. The capital city was named after the chief goddess of the Nephilim, Ashtaroth.

> Then we turned and went up the road to Bashan; and Og king of Bashan came out against us, he and all his people, to battle at Edrei. And the LORD said to me, "*Do not fear him*, for I have delivered him and all his people and his land into your hand; you shall do to him as you did to Sihon king of the Amorites, who dwelt at Heshbon."

> So the LORD our God also delivered into our hands Og king of Bashan, with all his people, and we attacked him until he had no survivors remaining. And we took all his cities at that time; there was not a city which we did not take from them: sixty cities, all the region of Argob, the kingdom of Og in Bashan. All these cities *were* fortified with high walls, gates, and bars, besides a great many rural towns. And we utterly destroyed them, as we did to Sihon king of Heshbon, utterly destroying the men, women, and children of every city. But all the livestock and the spoil of the cities we took as booty for ourselves.

> And at that time we took the land from the hand of the two kings of the Amorites who *were* on this side of the Jordan, from the River Arnon to Mount Hermon (the Sidonians call Hermon Sirion, and the Amorites call it Senir), all the cities of the plain, all Gilead, and all Bashan, as far as Salcah and Edrei, cities of the kingdom of Og in Bashan.

For only Og king of Bashan remained of the remnant of the
giants. Indeed his bedstead *was* an iron bedstead. (*Is* it not
in Rabbah of the people of Ammon?) Nine cubits *is* its length
and four cubits its width, according to the standard cubit.
—Deuteronomy 3: 1-11

Joshua and the Israelites were instructed to destroy these demonic
descendants, and the bloodlines of these giants, wherever they were found.
The fallen sons of God had crossed DNA boundaries that were not pleasing to Him. Therefore, Yahweh promised to remove their offspring from
the Earth.

Leveling the Battlefield and Using God's Wrath
The Lord told the Israelites to "fear not." Whenever God tells us not to
be afraid, we must trust in His battle strategies. The Lord strategically
drove King Og and the inhabitants of his sixty cities away from their high
walls and city gates and into an open battlefield. The Lord created a level
battlefield for the Israelites, so they could overthrow their enemies. "I sent
the hornet ahead of you, which drove them out before you—also the two
Amorite kings. You did not do it with your own sword and bow" (Joshua
24:12).

As you pray for your family, community, and nation, the Lord will
flush out your enemies, exposing them before you to be dismantled on
a level battlefield. With God on your side, their high fences and massive
walls of deceit will no longer protect them. He will fight against those
who fight against you. He will be your shield and buckler, and He will
stand up for you (Psalm 35:2).

You must take no chances with the evil that is displayed in your families, communities, businesses, and nations. Fight in the spiritual realm
through prayer and fasting. Even as these giants were no match for God
in ancient history, the spiritual giants of today are no match for Him
either.

The entire nation of the Canaanites quaked with fear at the arrival
of the Israelites in the promised land, and so shall your enemies quake

with fear in the spiritual realm for Jehovah Gibbor is with you. Your prayers and intercessions will be the instruments of God's wrath upon the kingdom of darkness. Remember, your very DNA can be encoded with his Bioglory, and His Life-Light can shine through you, helping you to overcome the darkness.

True followers of the Lord hate all manner of evil and darkness (Proverbs 8:13). The ancient disciples had a hatred for the demonic realm (Psalm 139:21-22). Their wrath and God's wrath worked together to defeat many enemies. They were able to reclaim the promised land.

Let's take a look at another example of how God's faithful followers used His wrath to defeat darkness. When Goliath came up against the Israelites to take their possessions, David stood up and dismantled the famous giant. Goliath was "a very tall man with six fingers on each hand and six toes on each foot—for a total of 24 digits—who was a descendant of the Rephaim" (1 Chronicles 20:6). But David was not alone in his victory. Another battle with the Philistines was in progress at Gob. At the same time, Sibbekai the Hushathite killed Saph, one of the descendants of Rapha. In another battle, Elhanan son of Jair killed the brother of Goliath. When David became King of Israel, his army killed many more giants and descendants of the fallen sky gods.

When the Philistines captured the Ark of the Covenant and placed it in the temple of Dagon in Ashdod, God destroyed their graven images. Two mornings in a row, the image of Dagon was found flat on its face before the Ark. The second time, Dagon's head and hands were cut off. The severing of Dagon's head and hands is symbolic. Yahweh dismantled Dagon's head (his power and authority) and hands (his actions or works). Dagon was a hybrid—his DNA was half man, half fish. He was a deity over the great deep, and his cosmic powers ruled over the Deep Ones, an amphibious humanoid race indigenous to Earth's oceans. God was not happy with the followers of Dagon for crossing DNA boundaries, and like all those hybrids before them, they were wiped from the earth.

Iron Mixed with Miry Clay
As you will remember, the giants were mortals with angelic strength and supernatural capabilities. The regulatory switches within their junk DNA

were encoded with the genetic language from the fallen angels who had cohabitated with human women. Some say that the giants were one-third human and two-thirds god. This point will become even more important and relevant as we learn more about our enemies in the end times.

Daniel 2:41-43 shows that species will mix and DNA will be manipulated in the end times as the enemy tries to restore Project Babel:

> Whereas you saw the feet and toes, partly of potter's clay and partly of iron, the kingdom shall be divided; yet the strength of the iron shall be in it, just as you saw the iron mixed with ceramic clay. And as the toes of the feet were partly of iron and partly of clay, so the kingdom shall be partly strong and partly fragile. As you saw iron mixed with ceramic clay, they will mingle with the seed of men; but they will not adhere to one another, just as iron does not mix with clay. In the days of those kings, the God of heaven will set up a kingdom that will never be destroyed, nor will it be left to another people. It will shatter all these kingdoms and bring them to an end, but will itself stand forever. And just as you saw a stone being cut out of the mountain without human hands, and it shattered the iron, bronze, clay, silver, and gold, so the great God has told the king what will happen in the future. The dream is true, and its interpretation is trustworthy.

The phrase "iron mixed with miry clay" is acknowledged by most scholars as the co-mingling of the fallen sons of God with the daughters of men. As we have documented, the stone being cut out of the mountain without human hands is Christ, the Son of God. The second coming of Christ will bring an end to the comingling of DNA.

What does the modern-day press say about genetic modifications? According to *Science Magazine*, He Jiankui, a Chinese researcher, stunned the world by announcing he had helped produce genetically modified babies. According to *Obstetric & Gynecologic Magazine,* Dr. Jiankui announced the delivery of twins Lulu and Nina, conceived with embryos that had been genetically modified using Clustered Regularly Interspaced

Short Palindromic Repeats (CRISPR-Cas9), at a gene editing conference in Hong Kong on November 26, 2018. This news was met with varying reactions. Some people marveled at the scientific advancements, while others saw that they were a defilement of God's law.

The existence of giants are referenced both in the Bible and in scientific magazines. The Bible makes approximately two hundred references to the giants (the illegitimate offspring of the fallen ones), and many modern-day researchers have also described the giant race:

- Scientific American – "Ancient American Giants" (August 14, 1880; pg. 106)
- Urbana Union – "Skeletons of a Giant Race Found near Petosi" (February 16, 1870; pg 1)
- The Vancouver Sun – "Primitive Man, Ten Feet Tall, Is Unearthed" (August 18, 1922; pg. 9)
- Hopkinsville-Kentuckian – "The Bones of a Giant Ten Feet in Height Found Near Lewisport" (April 23, 1897; pg. 8)
- The Times Dispatch – "Giant's Tooth Found" (February 11, 1907; pg. 8)

Additionally, North American natives, especially the Chippewa, Sandusky, Tawa, Iroquois, Cherokee, Choctaw, and Hopi tribes, believed a race of giant beings existed before them.

As we continue to look at the operations of Project Babel and the crossing of DNA boundaries, we see that the sky gods had a hold on many nations. When we face the kinds of evil cosmic power today that were prevalent in ancient times, we will need God's power for deliverance:

> Why do the nations rage, And the people plot a vain thing? The kings of the earth set themselves, And the rulers take counsel together, Against the LORD and against His Anointed, *saying*, "Let us break Their bonds in pieces And cast away Their cords from us."
> —Psalm 2:1-3

As today's church stands up against these evil cosmic powers, which take counsel against the Lord and against His Anointed, we need to stand in the victory of Christ, praying and speaking against the crossing of DNA boundaries. We are dealing with very serious issues in our communities and nations, and those issues will only grow until Christ comes again in the end times. The good news is that the kingdom of God is currently drawing all nations from every corner of the earth (Isaiah 2, Hebrew 2:22). We are building His army. The kingdoms of man are becoming the kingdoms of our Lord, and He will reign forever and ever (Revelation 11:15).

Upon This Rock

We previously discussed the famous story of Peter's proclamation of Jesus as the Christ Son of the Living God (Matthew 16:13-20). Soon after, Christ responded, "I tell you, you are Peter, and on this rock I will build my church, and the gate of Hades shall not prevail against it." This proclamation was made in the district of Caesarea Philippi, in the heart of the rocky Terence of Bashan (known as the place of Azazel), which sits at the foothills of the cosmic mountain known as Mount of Hermon.

This location is significant for many reasons. First, it was known as the "gate of Hades," the main entrance to the underworld and the abode of dead souls. Second, Azazel is known as the scapegoat in Leviticus 16: 8-10. Aaron had to cast two lots, "one lot for the Lord and the other lot for the scapegoat. And Aaron shall bring the goat on which the Lord's lot fell, and offer it as a sin offering. But the goat on which the lot fell to be the scapegoat shall be presented alive before the Lord, to make atonement upon it, and to let it go as the scapegoat into the wilderness."

According to some biblical scholars, all demonic entities are banished into "dry places"—empty, waste lands, where little or no human beings are found. Modern occultism says that Azazel's goat is Baphomet, and many today worship him as the "desert demon" who resides on the cosmic Mount of Hermon.

Finally, some biblical scholars point out that "this is a ground zero for the fallen gods against whom Christ and His church is fighting. The cosmic Mount of Hermon is known as the location where the fallen sons of God first came to earth.

When Jesus told Peter, "upon this rock I will build my church," it was a prophetic implication that His church will literally be "seated above" ALL the cosmic power of the gate of Hades.

In Matthew 16:18, Christ establishes His church upon the rock, indicating that His church body is the DNA ladder and gateway between heaven and earth. The Father encodes His power and authority through this gateway. In addition, while at the gates of Hades, Jesus was fortifying the Lord's kingdom and declaring war on Hades.

Then He immediately took His disciples, Peter, James and John, to the top of the rock (Mount Hermon) to demonstrate exactly how the church would encode God's power through prayer. "And as He [Jesus] was praying, the fashion of his countenance was altered, and his raiment was white and glistering" (Matthew 17:2, Mark 9:2, Luke 9:29).

Through prayer, we have the power to store, replicate, and manifest the Life-Light of God on earth. Through prayer-encoding Bioglory, we are transfigured. And the gates of hell cannot prevail against us because Christ in the flesh has created Bioglory Morphic Resonance Fields for the church. We have been morphed into the image of Christ in the flesh. Our bodies are the gateway of God.

As Jesus was demonstrating the encoding power of prayer, Moses and Elijah appeared. They were representations of the Old Covenant, summed up as the Law (Moses) and the Prophets (Elijah). Christ in the flesh is the king of the New Covenant that both the Law and Prophets pointed toward.

Christ has torn down the gates of Hades upon the cosmic Mount of Hermon, and He celebrated His victory by establishing His power on a new cosmic mountain, Zion.

> He that sitteth in the heavens shall laugh: the LORD shall have them in derision. Then shall he speak unto them in his wrath, and vex them in his sore displeasure. Yet have I set my king upon my holy hill of Zion. I will declare the decree: the LORD hath said unto me, Thou art my Son; this

day have I begotten thee. Ask of me, and I shall give thee the heathen for thine inheritance, and the uttermost parts of the earth for thy possession.
—Psalm 2:4-8

David also proclaimed the victory of God over the cosmic Mountain of Bashan, which was a foreshadowing of Christ's proclamation to Peter about the church:

A mountain of God is the mountain of Bashan; A mountain of many peaks is the mountain of Bashan. Why do you fume with envy, you mountains of many peaks? This is the mountain which God desires to dwell in; Yes, the Lord will dwell in it forever. The chariots of God are twenty thousand, Even thousands of thousands; The Lord is among them as in Sinai, in the Holy Place. You have ascended on high, You have led captivity captive; You have received gifts among men, Even from the rebellious, That the Lord God might dwell there. Blessed be the Lord, Who daily loads us with benefits, The God of our salvation! Selah Our God is the God of salvation; And to God the Lord belong escapes from death. But God will wound the head of His enemies, The hairy scalp of the one who still goes on in his trespasses. The Lord said, "I will bring back from Bashan, I will bring them back from the depths of the sea, That your foot may crush them in blood, And the tongues of your dogs may have their portion from your enemies."
—Psalm 68: 15-22

The Gates of Hades

One last passage ties into this revelation of how Christ has established His church as His DNA ladder and gateway, and the gates of Hades cannot prevail against it. Have you ever wondered why the Lord said, "I Loved Jacob, but Esau I Hated." Well, I have, and it is mind-blowing to see how it all ties together.

This is a divine revelation. The Lord spoke his word to Israel through Malachi. "I loved you," says the Lord. "But you ask, 'How did you love us?' Wasn't Esau Jacob's brother?" declares the Lord. "I loved Jacob, but Esau I hated. I turned his mountains into a wasteland and left his inheritance to the jackals in the desert." The descendants of Esau may say, "We have been beaten down, but we will rebuild the ruins." Yet, this is what the Lord of Armies says: "They may rebuild, but I will tear it down. They will be called 'the Wicked Land' and 'the people with whom the Lord is always angry.' You will see these things with your own eyes and say, 'Even outside the borders of Israel the Lord is great.'"
—Malachi 1:1-5

God hated Esau because he worshiped sky gods. His DNA was profaned and offered as a gateway for the fallen sons of God. Esau turned away from God to worship the goat demons on the cosmic mountain; therefore, the Lord specifically said, "but Esau I hated, I *turned his mountains* into a wasteland and left *his inheritance* to the jackals in the desert." Esau's mountain represents his pagan altar and place of worship, so God turned it into a wasteland. (Remember that altars and DNA share similar qualities because they are both gateways.)

Esau Despised His Birthright
Do you remember the legal contract that Jacob negotiated with Easu? He offered to give Esau a bowl of stew in exchange for his birthright, and Esau agreed because he despised his birthright.

The birthright (bekorah) has to do with both position in society and DNA inheritance. If Esau had received the birthright blessing from Isaac, then Abraham's bloodline would have been profaned because Esau's DNA was corrupted by idolatry. Jacob instead received the birthright and carried on the unblemished bloodline through which the Messiah, Christ, would be born. We can connect this spiritual truth to what we have previously studied about Jacob's DNA ladder, a gateway through which God encoded His power and glory.

In Obadiah 1:15-18, the Prophet proclaimed:

> But on Mount Zion there shall be deliverance, And there shall be holiness; The house of Jacob shall possess their possessions. The house of Jacob shall be a fire, And the house of Joseph a flame; But the house of Esau shall be stubble; They shall kindle them and devour them, And no survivor shall remain of the house of Esau," For the LORD has spoken. The South shall possess the mountains of Esau, And the Lowland shall possess Philistia. They shall possess the fields of Ephraim And the fields of Samaria. Benjamin shall possess Gilead.

God will not leave us ignorant concerning the schemes of the devil. Whenever Satan has tried to profane the bloodline, the Spirit of the Lord has lifted up a standard against him.

Now we have a more complete understanding of why God hated Esau but loved Jacob. God cannot encode His Bioglory Life-Light into any temple (body) that has been given over to idolatry. The sin of idolatry is connected to the crossing of DNA boundaries and the giving up of the body, soul, and spirit to the worship of the fallen sons of God. Second Corinthians 6:16-17 says, "What agreement has the temple of God with idols? For we are the temple of the living God; just as God said, 'I will dwell in them and walk among them; And I will be their God, and they shall be My people. Therefore, come out from their midst and be separate,' says the Lord. 'And do not touch what is unclean; And I will welcome you.'"

The Great Commission

God has a remnant, a people that are set aside for His Glory. He has given us the authority to trample on snakes and scorpions (those who have crossed the DNA boundary) and to overcome all the power of the enemy.

In Matthew 10, Jesus commissioned His disciples to go out and minister unto the people. He knew they would come face-to-face with these Nephilim spirits, and He said to them, "I give you power," and this power was the ability to cast out unclean spirits. We also read about this power in the book of Luke:

> Then the seventy returned with joy, saying, "Lord, even the demons are subject to us in Your name." And He said to them, "I saw Satan fall like lightning from heaven. Behold, I give you the authority to trample on serpents and scorpions, and over all the power of the enemy, and nothing shall by any means hurt you. Nevertheless do not rejoice in this, that the spirits are subject to you, but rather rejoice because your names are written in heaven."
> —Luke 10:17-20

Again, Jesus gave the power and authority to do ministry work to all His disciples. No grand event took place for this commission to be valid. They were smack in the middle of a spiritual war, and Jesus said to "go and cast them out, and to heal all kinds of sickness and all kinds of disease" (Matthew 10:1).

So why are so many believers disregarding and denying the commission to go out and wage war on these demonic forces? I believe that many are waiting on someone else to go out and do God's work. But Jesus has already given the commission to the church, and we don't need to wait for a grand event: "He said unto them, Go ye into all the world, and preach the Gospel to every creature . . . And these signs shall follow them that believe; In my name shall they cast out devils" (Mark 16: 15-20). The preaching of the gospel is a testimony to the nations that Christ in the Flesh has defeated the fallen sons of God and their illegitimate offspring (Matthew 24:14). The gospel is to be proclaimed to all creation under heaven (Colossians 1:23).

Who are these devils that we are supposed to cast out? The demonic forces that the church is fighting against are the spirits of the Nephilim (giants) that died in the flood. God drowned them. Because these Nephilim were born of human women, they had physical bodies. After their bodies perished in the flood, their evil spirits began roaming the earth. They are now disembodied, unclean spirits roaming the dry places of the earth until Judgment Day.

The legion of demons said to Jesus, "What have we to do with You, Jesus, You Son of God? Have You come here to torment us before the time?" (Matthew 8:29). The demons were asking if Jesus had come to pass final judgment upon them before the end times. Throughout Jesus's ministry, the disembodied demons would often ask Him this question because they knew that their day of final judgment was fast approaching.

Demons do not have physical bodies. In the Old Testament, they are referred to as "familiar spirits." In the New Testament, they are called "unclean, foul spirits." You are God's temple, filled with His Holy Spirit, and all disembodied spirits are illegal in the earth realm; therefore, you have power over them (Luke 10:19; Matthew 10:1; Eph 1:20; 2:6).

Many believers suffer unnecessarily because they ignore the demonic realm and fail to exercise their authority over it. Do you want to see changes in your region, city, and nation? You are a king adorned in the blooded priestly garment of Christ, and you have the power to command every evil spirit in any geographical region. As Ecclesiastes 8:4 says, "Where the word of a king is, there is power; And who may say to him, 'What are you doing?'"

You are a link in the chain of the bloodline of Christ, and you have His power. The Bible says that we are a kingdom of kings and priests on the earth, and we have been given dominion over the earth (Revelation 5:10). So, with that being said, you have the ability to pray and command things to come into alignment with the Word of God.

During His earthly ministry, Jesus exercised authority over every disembodied spirit. When Jesus and His disciples were in a Gentile region, they came upon a man possessed with many evil spirits. The spirits made him supernaturally strong and uncontrollable "because he had often been bound with shackles and chains. And the chains had been pulled apart by him, and the shackles broken in pieces; neither could anyone tame him" (Mark 5:4).

> When he saw Jesus from afar, he ran and worshipped Him.
> And he cried out with a loud voice and said, "What have

I to do with You, Jesus, Son of the Most High God? I implore You by God that You do not torment me." For He said to him, "Come out of the man, unclean spirit!" Then He asked him, "What *is* your name?" And he answered, saying, "My name *is* Legion; for we are many." Also he begged Him earnestly that He would not send them out of the country. Now a large herd of swine was feeding there near the mountains. So all the demons begged Him, saying, "Send us to the swine, that we may enter them." And at once Jesus gave them permission. Then the unclean spirits went out and entered the swine (there were about two thousand); and the herd ran violently down the steep place into the sea, and drowned in the sea.
—Mark 5:6-13

A Body is Needed

These disembodied spirits want to dwell on earth, but that is impossible without a physical body. Thus, they would rather dwell in the pigs than be cast into the wilderness or the dry places of the earth. Disembodied spirits must obey Christ and the children of God because we have legal rights to exercise His authority over them in the earth realm.

Nephilim spirits are familiar with the fallen nature of men and the crossing of DNA boundaries. Therefore, familiar spirits are the most instrumental spirits in the kingdom of darkness. Because demons are disembodied spirits, they can possess a human or animal, or attach themselves to inanimate objects (charms, amulets, and so forth). They must attach themselves to something they're familiar with to remain relevant in the earth realm. This is why they're also called "familiars." They are acquainted with corporeal bodies and can use them as a covering or house for their spirit.

When an unclean spirit goes out of a man, he goes through dry places, seeking rest, and finds none. Then he says, "I will return to my house from which I came." And when he comes, he finds it empty, swept, and put in order. Then he goes and takes with him seven other spirits more wicked than himself, and they enter and dwell there; and the last state of that man is worse than the first. So shall it also be with this wicked generation.

—Matthew 12:43-45

The Hebrew word used in the Bible to describe a familiar spirit is *obe*, which translates to "leather bottle." Leather bottles were designed to be folded into familiar shapes and carried in a way that made them difficult to find. Similarly, familiar spirits come to hide themselves in our bodies and create attachments to our souls. These attachments are called *soul ties*.

Like the leather bottles, these soul ties can conform to odd shapes to hide themselves, going almost undetected. When people repetitively sin or harbor evil thought patterns, these spirits can delve deeper and deeper into the intimate portions of the soul. Familiar spirits can commune with people through vain imaginations, thoughts, dreams, inherited genetic traits, or generational curses.

Mediums and fortunetellers are familiar with and connected to the realm of the dead. The Mosaic Law of the Old Testament prescribed the death penalty for anyone who consulted with familiar spirits: "And the person who turns to mediums and familiar spirits, to prostitute himself with them, I will set My face against that person and cut him off from his people . . . A man or a woman who is a medium, or who has familiar spirits, shall surely be put to death; they shall stone them with stones. Their blood shall be upon them" (Leviticus 20: 6, 27).

King Saul rebelled against God's command by seeking guidance from a woman with a familiar spirit. Disguising himself, Saul pleaded with the witch to call up the spirit of the deceased prophet Samuel to devise a battle plan against the Philistines (1 Samuel 28:7-8). God put a strict ban on consulting demons, and Saul ultimately died for this sin:

> So Saul died for his unfaithfulness which he had commit-
> ted against the Lord, because he did not keep the word
> of the Lord, and also because he consulted a medium for
> guidance. But he did not inquire of the Lord; therefore He

killed him, and turned the kingdom over to David the son
of Jesse.
—1 Chronicles 10:13-14

Earlier, we examined the biblical principle that the sinful nature of
Adam was passed down to all humans, with the exception of Jesus Christ.
Logically speaking, we must assume that the fallen angels also passed their
spiritual nature on to the Nephilim, their demon offspring. As hybrids, the
Nephilim constituted an entirely new race with genes from both humans
and angels. Therefore, they had a different type of spirit altogether.

People often use demons for occult practices and medical practices. They access the spirit realm by summoning the demons to do their
bidding:

> And it came to pass, as we went to prayer, a certain damsel
> possessed with a spirit of divination met us, which brought
> her masters much gain by soothsaying. This girl followed
> Paul and us, and cried out, saying, "These men are the ser-
> vants of the Most High God, who proclaim to us the way of
> salvation." And this she did for many days.
> —Acts 16:16-18

Demons will attach themselves to downcast people. The demons will
then forcefully perpetuate cycles of failure, sin, and death within those
people's lives. When there is unbelief in God, familiar spirits lodge into
our minds and drive us to false idol worship. Paul said:

> Rather, we have renounced secret [hidden] and shameful
> [familiar] ways; we do not use deception, nor do we distort
> the word of God . . . The god of this age has blinded the
> minds of unbelievers, so that they cannot see the light of
> the gospel of the glory of Christ, who is the image of God.
> —2 Corinthians 4:4

The Bible tells us to pull down every suggestion, projection, or lofty thought that seeks to exalt itself above God. These suggestions come from familiar spirits, who are whispering in our ears.

How do we recognize familiar spirits? The only way I began to recognize them was by challenging the voices I heard in my mind. Demons mutter and speak from the spiritual realm. But their voices are not the voice of the Holy Spirit. *You will know the difference by what they speak.* The Holy Spirit will never bring bad intentions to your mind. If you're receiving evil or carnal thoughts, you should ask the Holy Spirit to show you any open gate or doorway to your DNA, mind, or soul that is giving these familiar spirits access to you. Then you need to close that pathway by praying the Word. This will dismantle the hidden demonic forces seeking to destroy your life.

The tremendous importance of these points of entry to your DNA, mind, and soul explains why Jeremiah lamented over the desolated city gates:

> The roads to Zion mourn Because no one comes to the set feasts. All her gates are desolate; Her priests sigh, Her virgins are afflicted, And she *is* in bitterness
> —Lamentations 1:4

Securing Your Gates in Prayer

You must speak strategically and directly to the desolated gates and command them to close to demonic counsel. The gates to your body are the ear gate (hearing), eye gate (sight), mouth gate (taste), nose gate (smell), feel gate (touch), and DNA gate (genetic inheritance).

Assign angels of the Lord to guard your gates. Do not allow these demonic spirits to speak information to your soul's realm—your mind, imagination, reason, conscience, and emotion. Cast them out by praying the Word of God. "My sheep knows my voice and another they will not entertain" (John 10:5).

These disembodied spirits will gain access to your soul to rob you of your life. "Most assuredly, I say to you, he who does not enter the sheepfold

by the door, but climbs up some other way, the same is a thief and robber . .
. The thief does not come except to steal, and to kill, and to destroy. I have
come that they may have life, and that they may have it more abundantly"
(John 10: 1-5, 10).

God is gathering anointed prayer warriors filled with the Holy Spirit
and giving them jurisdictional authority over these disembodied spirits, so
that families, communities, and nations can be brought back into divine
alignment. This alignment is to be accomplished by praying the will of
God and imposing the mandate of the kingdom of heaven over and above
the mandate of the kingdom of darkness. "Thy Kingdom come, They will
be done, on earth as it is in heaven."

As a king on the earth, your faith-filled words are shaping and
forming things into what the Father desires to see on earth. Atoms, molecules, crystal, cells, dimensions, and realms are recalibrating. You have the
ability to realign DNA to reflect His Bioglory.

Christ Is Not a Disembodied Spirit
Christ defeated Satan through the cross! He died and was resurrected in
full bodily form. Christ is not a disembodied spirit. To prove He was truly
resurrected to a glorified body, Jesus said to His disciples, "touch me":

> But they were terrified and frightened, and supposed they
> had seen a spirit. And He said to them, "Why are you trou-
> bled? And why do doubts arise in your hearts? Behold My
> hands and My feet, that it is I Myself. Handle Me and see,
> for a spirit does not have flesh and bones as you see I have."
> —Luke 24: 37-39

The fact that the disciples could see and touch Jesus proves that
He was not a disembodied spirit. Jesus dematerialized Himself, passed
through a wall, and then materialized His body again in front of the disciples. After His resurrection, Christ maintained His human nature on a
superior, immortal level.

During Christ's ministry, an astonishing amount of demonic activity
occurred on earth. These disembodied spirits had a number of encounters

with the Lord, and they always knew His true identity. Consider these passages:

> And the unclean spirits, whenever they saw Him, fell down before Him and cried out, saying, "You are the Son of God." But He sternly warned them that they should not make Him known.
> —Mark 3:11-12

> Now in the synagogue there was a man who had a spirit of an unclean demon. And he cried out with a loud voice, saying, "Let us alone! What have we to do with You, Jesus of Nazareth? Did You come to destroy us? I know who You are—the Holy One of God!" But Jesus rebuked him, saying, "Be quiet, and come out of him!" And when the demon had thrown him in their midst, it came out of him and did not hurt him. Then they were all amazed and spoke among themselves, saying, "What a word this is! For with authority and power He commands the unclean spirits, and they come out." And the report about Him went out into every place in the surrounding region."
> —Luke 4:33-37

These passages reveal a lot about the relationship between demons and the Lord, Christ Jesus. The inhabitants of the kingdom of darkness are aware of the true identity of Christ. They referred to Jesus as the "Holy One"—a title that signifies Him as both God and Messiah (man).

The demons are under the authority of Christ, and He has given the church the same authority. We have the power to subdue these dark forces (Genesis, 1:28; Luke 10:19). Every time a believer casts out a demon, it is a testimony of the sovereign power of God over all creation and a testimony of the connection we share with Him.

The body of Christ is the enemy of all disembodied spirits. Everything the Nephilim do is contrary to God's mandate that man should have dominion over the earth. God said, "Let them [man] have dominion."

God didn't say, "Let spirit beings have dominion over the earth." In the beginning, God removed all spirit beings and gave the earth to man (spirit, soul, and body) to rule. To have dominion over the earth, these three things must be combined in one entity. God's Word is also a Law unto Himself. He keeps His Word and watches over it.

In the end times, when Christ in the flesh arrives to reestablish the kingdom of God, He will do so by "binding the strong man," who is known as Satan or "the god of this world." He will also cast out all disembodied demon spirits. During his ministry, when Christ cast out demons, it was foreshadowing of Him casting down Satan's power and standing in dominion over the earth (John 12:31). He destroyed the one who has the power of death, that is the devil (Hebrew 2:14).

Jude 1:25 proclaims: "To the only wise God our Saviour, be glory and majesty, dominion and power, both now and ever. Amen."

Let's recall Jesus's astonishing statement, "All authority in heaven and on earth has been given to me" (Matthew 28:18). Now let's take a look at the anointing of our conquering King who lives within us. Christ is in you. We are now executing all that has been written about the DNA of Christ. This honor to do His work is given to the church. You have authority from the King to advance His kingdom over all the earth. The Lord says, "Ask me, and I will make the nations your inheritance, the ends of the earth your possession" (Psalm 2:8).

King David understood the importance of prayer before and after victory. He fasted and prayed for the defeat of his enemies, and God answered him. He had many victories in both the physical and spiritual realm, and he expressed gratitude for his people's deliverance through prayer. Like David, we will prevail over our enemies as we bring down every demonic entity. Like David, we must be thankful for God's aid.

It is my prayer that you spend countless hours in the presence of God to take in all that He has made available to you through His Bioglory. David would often end his prayers by asking Him to fill the whole earth "with the glory of God." I truly believe that the earth will be filled with the fullness of God through prayer-encoding Bioglory. Just imagine: every plant, animal, human, and living organism's DNA will start absorbing, storing, and emitting the Life-Light of God in the end times. What a beautiful sight this will be.

> The heavens declare the glory of God; and the firmament sheweth his handywork. Day unto day uttereth speech, and night unto night sheweth knowledge. There is no speech nor language, where their voice is not heard. Their line is gone out through all the earth, and their words to the end of the world. In them hath he set a tabernacle for the sun, Which is as a bridegroom coming out of his chamber, and rejoiceth as a strong man to run a race. His going forth is from the end of the heaven, and his circuit unto the ends of it: and there is nothing hid from the heat thereof . . . The fear of the LORD is clean, enduring for ever: the judgments of the LORD are true and righteous altogether.
> —Psalm 19:1-9

Applying His Message

Preacher and author Dr. Bill Winston said, "Until we have a revelation of our right to execute vengeance, we will remain tormented by wicked forces." Prayer is one area where believers struggle to apply what they know. When they are given a good revelation about prayer, they decide to be consistently prayerful, but many never follow through. They stop praying and believing what the Word has revealed to them after a few days.

Whenever you receive a revelation from the Holy Spirit, you must quickly apply it to your life. Breakthroughs come when revelation is applied, for applied revelation is one of the keys that unlocks God's authority on the earth. The keys of the kingdom were only given to the church after Peter applied (spoke) the revelation that was given to him by the Father. Christ is the anointed King, ruling from the cosmic mountain of Zion in the city of God. We are His church positioned on earth but seated in Christ, far above all principalities and powers. As we pray, we are standing in the victory of our Lord and applying divine revelation to our lives.

Jesus reigns victoriously, and He is always praying for us. He undid Project Babel, reprogrammed the human genome system, and gave the Father full access to humanity through the DNA Ladder (Acts 2; Colossians 2:9). He has begun to draw nations away from the Tower of Babel and their sky gods and draw them toward His cosmic mountain, Mount Zion, instead (Isaiah 2). This deliberate act is also known as the "War of the Seed." Christ is putting an end to the crossing of DNA boundaries.

> Then cometh the end, when he shall have delivered up the kingdom to God, even the Father; when he shall have put down all rule and all authority and power. For he must reign, till he hath put all enemies under his feet. The last enemy that shall be destroyed is death. For he hath put all things under his feet. But when he saith all things are put under him, it is manifest that he is excepted, which did put all things under him. And when all things shall be subdued unto him, then shall the Son also himself be subject unto him that put all things under him, that God may be all in all.
> —1 Corinthians 15:24-28

CHAPTER 11
ULTIMATE FORCE:
CALLING SYSTEMS INTO ALIGNMENT

The appearance of the wheels and their work was like unto
the colour of a beryl: and they four had one likeness: and
their appearance and their work was as it were a wheel in
the middle of a wheel.
—Ezekiel 1:16

When reading the word "wheels" or the phase "the spirit of the living
creature was in the wheels," most readers immediately conjure images
of a physical wheel like a car tire or some other wheel. I, on the other
hand, think of an atom, turning as a wheel, going around and around, with
energy of electrons, protons, and neutrons that are the color of a beryl.

Every system within the universe is composed of atoms, whether
that system is a living organism, a spiritual being, or an inanimate object.
Atoms are the building blocks and defining structures of all elements. In
fact, *you*, as a person, are also a system composed of atoms.

A single atom consists of a strong nuclear force that holds the quarks,
protons, neutrons, and electrons together. This *strong nuclear force* is one
of the four fundamental forces in nature. The other three are gravity,
electromagnetism, and the weak force. As its name implies, the strong
nuclear force is the *strongest* force of the four. It is responsible for binding
together the fundamental particles of matter to form the larger systems
that we see with our physical eyes. Wherever a strong nuclear force or
wheel of the atom flows, particles are bound and matter is created.

If the strong nuclear force is the strongest force in the universe, then
the force that created it must be the Ultimate Nuclear Force. And that
Ultimate Nuclear Force is the Word of God. All things were made by
Him. He created life. That life is the light of men. In other words, the
Ultimate Nuclear Force is made of His Bioglory and Life-Light.

Ezekiel's Wheel

When we look at Ezekiel's mysterious wheel, we can see that it represents the atom. The atom is the power of the Holy Spirit that flows from the throne of God throughout all creation, holding things together by means of the Ultimate Nuclear Force. "And when the living creatures went, the wheels went by them: and when the living creatures were lifted up from the earth, the wheels were lifted up. Whithersoever the spirit was to go, they went, thither was their spirit to go; and the wheels were lifted up over against them: for the spirit of the living creature was in the wheels" (Ezekiel 1:19-20).

The Holy Spirit within the Ultimate Nuclear Force can bring all systems into alignment through prayer. Let's dig into this idea further.

- Can the strong nuclear force that holds the particles of cancerous cells together be dematerialized by the Ultimate Nuclear Force, never to materialize again?
- Can the strong nuclear force that holds *together* the particles of demonic kingdoms, *systems, and realms be loosed and recalibrated by* the Ultimate Nuclear Force?
- Can the strong nuclear force that holds the particles of a broken marriage together take on a new form of life and begin to blossom again through the power of the Ultimate Nuclear Force?

Yes, the Ultimate Nuclear Force can do all of these things, and more! The nucleus is the central and most important part of any object, movement, group, or system. It forms the basis for intrinsic and extrinsic activity, growth, and power. All information is registered within the nucleus. Every interaction in the universe is a result of how the information within the nucleus is altered. Therefore, when you pray the Word of God, you are sending information from the Ultimate Nuclear Force and speaking directly into the nucleuses of your atoms to alter your day. Romans 1:20 reads, "For the invisible things of him from the creation of the world are clearly seen, being understood by the things that are made, even his eternal power and Godhead; so that they are without excuse." In this scripture,

God is saying that we are responsible for bringing alignment to these systems—no excuses. In this chapter, we will look at how to bring our own bodies, words, and deeds into alignment with God. We will also examine how we can align other systems through God's Bioglory.

No matter what you do during the day, your body and mind must be recalibrated and reconciled with the eternal power of the Godhead consistently. One way to recalibrate yourself is by kneeling in prayer every night. Every proton, neutron, and electron that has been misaligned by the enemy throughout the day can then come into divine alignment. "When You send Your Spirit, they are created, and You renew the face of the earth. May the glory of the LORD endure forever; may the LORD rejoice in His works, He looks on the earth, and it trembles; He touches the mountains, and they smolder" (Psalm 104: 30-32).

When the Holy Spirit moves, He invades every atom within all the stratospheres, atmospheres, hemispheres, regions, realms, and domains, creating a BM-Field so miracles can occur. The movement of the Holy Spirit alters divine systems, renewing the face of the earth and making all things come together according to His will (Isaiah 43:19; 65:17; Ephesians 2:15).

When you pray and speak to the strong nuclear force of your situation, you are disallowing every diabolical sanction, injunction, mandate, or order that opposes the will of the Lord concerning your life.

> Assuredly, I say to you, whatever you bind on earth will be bound in heaven, and whatever you loose on earth will be loosed in heaven.
> —Matthew 18:18

> Whatever things you ask when you pray, believe that you receive them, and you will have them.
> —Mark 11:24

The Power of the Word
Every day of your life, you are transmitting energy or information from the Ultimate Nuclear Force (Word of God) into every atom to create an

atmosphere around you. You were created by God—endowed with great purpose and power—and you have the mind of Christ to order systems at all levels of complexity to know their place (2 Peter 1:4; 1 Corinthians 2:16). With our faith-filled words, we can shape and form our days, weeks, months, and years ahead. When you use the Word of God to speak into the atmosphere, or to any given situation, that energy from the Word of God goes into the atoms and transforms them—just like the fig tree that dematerialized on a molecular level when Jesus spoke to it.

This prayer strategy of speaking directly to the atoms is designed to shake evil out of illegal places, bringing systems into alignment with the Word at the quantum realm. If you have noticed lately that things seem different, that the immediate atmosphere around you has changed, this is because God is simply answering your prayers. His Bioglory Life-Light is invading your life, transforming you along with your situation.

Transformed by Beholding

Paul foretold that we would behold the glory of the Lord with unveiled faces and that we would be transformed into the same image. Let's look at what it really means to behold the glory of God, Who emits intense Bioglory.

We previously said that imaginations are thoughts in picture form, carried by neurons through electrochemical processes throughout the body; thus, we become what we behold. The implications of such a possibility can be wonderfully important; cells can behold light as a blueprint image.

There are two ways to behold the glory of Christ: faith and sight. These two principles are referred to throughout the scriptures multiples times. Your faith refers to your faith in the works of Christ manifested in this world. Your sight refers to your spiritual sight, which beholds Christ's Bioglory. His Bioglory is too bright, illustrious, and marvelous for our physical eyes.

The scriptures make it clear that both humans and angels cover their faces in the presence of God, which fits well with the Bible's description of God's Bioglory as a blinding light (Matthew 17:1–3; Acts

9:1–9; Revelation 1:16). We cannot look directly upon Him. So how do we behold Him by sight?

60 Trillion Eyes

The cell's nucleus captures genetic codes. In a way, it is like an eye. Because we have over 60 trillion cells in our bodies, we have over 60 trillion eyes that can gaze on Christ! So, every fiber of your being can be focused on and consumed by His Bioglory.

The Bible explains this phenomenon when Ezekiel describes God on His throne (Ezekiel 1:18; 10:12). He is extremely detailed in his prophetic writings; there are creatures said to have "eyes all over, front and back, their hands and their wings, are completely full of eyes, as were their four wheels."

And again, this phenomenon is mentioned in Revelation 4:8: "And each of the four living creatures had six wings and was covered with eyes all around and within. Day and night they never stop beholding God while saying: 'Holy, Holy, Holy, is the Lord God Almighty, Who was and is and is to come!'"

Most scholars say that the many eyes in these scriptures symbolize the four living creatures' eternal wisdom and knowledge of Christ. I also believe that the many eyes symbolize their ability to behold the intense Bioglory of God that emanates from every angle of His throne. We, too, are beholding His intense Bioglory with over 60 trillion eyes (cells) while we are being transformed into His exact image. We become what we behold.

DNA Efficiency

As we continue to behold Christ, our cells are listening for the Bioglory language to relay to the DNA within their nucleuses. DNA then responds by activating the necessary molecular switches. The language tells the DNA which gene ability or weakness to switch on and which gene ability or weakness to switch off. The process of seeing and responding to these coding instructions goes on in all 60 trillion cells all the time, day and night.

What is even more amazing is that when we are beholding Christ, our DNA replication process becomes more efficient. DNA efficiency is a measure of the degree to which our DNA listens and communicates with each other. As God's Bioglory activates more DNA, a higher state of spiritual consciousness takes place. This heightened state increases our coherence levels, and thus, the spirit man is quickened. When DNA starts to function at a higher efficiency, we become spiritually awake (Ephesians 5:14, Isaiah 52:1).

When Adam sinned, we truly devolved as a species, away from the presence of God. Presently, most people's DNA is functioning at a lower efficiency—the exception being those who are spiritually alive in Christ. This means that a large portion of humanity's DNA is not functioning by listening and responding. These people are spiritually blind and deaf. When DNA is functioning at 100 percent, we have the empowerment of the Holy Spirit fully manifested within our beings.

The DNA efficiency factor is not chemical, but informational. Therefore, incoherent communication between our spirit man and our DNA is the only thing that stands between us and perfect heath. Our cells and our bodies have lost the genetic language of Bioglory because of Adam's sin. We are vulnerable to disease because we (as a species) have not spent enough time beholding the glory of Christ. Christ is the light of men. His light carries information. To reject Christ is to reject Life-Light information and decrease DNA efficiency. When our DNA is functioning at lower levels, the process of gene-encoding malfunctions because it has no good information to download.

Listening and Responding

Now let's talk about sound. How important is sound to our existence? How powerful is the influence of sound? John 1:1 states, "In the beginning was the word, and the word was with God, and the word was God," and the scripture goes on to describe God as bringing the world into existence using speech (sound).

Bible science describes the origins of creation as being attributed to the use of pure sound. When the scientific community was working on The Human Genome Project, they found that language and syntax existed within our DNA and that it could be transformed and re-encoded by frequencies, sounds, or even worded information within a laser-light beam.

Grazyna Fosar and Franz Bludorf are Russian scientists who have successfully transformed informational patterns from one set of DNA into another. They transformed frog embryos into salamander embryos without lifting a scalpel or marking one incision. How? They were able to modify amphibian DNA simply by manipulating specific sound frequencies. They concluded that human DNA can be reprogrammed by speech and sound when specific frequencies are used.

Is it a stretch then to assume the Bioglory waves of God have the ability to transfigure the human genome, from one degree of glory to another?

Like light, sound carries information and many of the ancient rulers used sound to mortify or reprogram DNA:

> Therefore, as soon as they heard the sound of the horn, flute, zither, lyre, harp and all kinds of music, all the nations and peoples of every language fell down and worshipped the image of gold that King Nebuchadnezzar had set up.
> —Daniel 3:7

Without lifting a single sword, Nebuchadnezzar had successfully manipulated specific sound frequencies to reprogram DNA. As soon as they heard the sound, people of every language fell down and worshipped the false god.

Sound has a power all its own. Sounds, vibrations, and frequencies carry information. This is why music has such a power to draw us in and change our states of mind. How many of you have noticed that your mood can be affected by another person's voice or a particular type of music? Some sounds, vibrations, and frequencies can sway your mood.

As we have already concluded, DNA is not immutable. It can be transformed. Human cells use DNA like a radio uses an antenna, to receive vibrations and frequencies in order to tune the body.

Children are taught to say, "Sticks and stones may break my bones but words will never hurt me." However, nothing could be further from the truth. Spoken words can carry negative or positive frequencies with the potential to reprogram our DNA and ultimately shape our destiny in the realms of the mind, soul, and body. The flow of word frequencies is indeed powerful. Even the words of inanimate objects like your TV, cell phone, and computer cannot be dismissed as harmless.

One night I feel asleep watching the news. I thought I was having a nightmare, but I woke up to *The Exorcist*, at the part of the movie where the demon possessing the girl was speaking. This scene immediately resonated fear within me. Right away, I prayed out loud and recited the scripture, "For God has not given us a spirit of fear, but of power and of love and of a sound mind" (2 Timothy 1:7). Praying the Word out loud reprogrammed my DNA, and my fear was diminished.

We must examine every influence on our DNA. My suggestion would be to stay away from lustful, violent, demonic, or avaricious television shows, movies, and literature. Remember, your environment has a huge impact on your mind, spirit, and body.

Pay close attention to your natural or artificial speech paradigm. The cells in your body react to every silent and verbal word. Negativity will bring down your energy, and soon after, your immune system. In Psalm 42:5, David said, "Why are you cast down, O my soul? And why are you disturbed in me?"

The Tone of Your Thoughts

Thoughts are silent words. Your body responds to your thoughts, and your soul hears everything you think. It's time you start paying attention to your habitual, silent speech patterns. This examination is another way to bring your system into alignment with the will of the Lord. If you find anything amiss, you can pray to change your thought pattern to reflect a pattern more pleasing to the Lord.

Every time you speak, take a moment to listen to what you are actually saying. The tone of your thoughts create a positive or negative flow of energy into your DNA. Paul says:

> Finally, brethren, whatsoever things are true, whatsoever things are honourable, whatsoever things are just, whatso-

ever things are pure, whatsoever things are lovely, whatsoever things are of good report; if there be any virtue, and if
there be any praise, think on these things.

—Philippians 4:8

Linguistics is the study of human speech. Linguists study the structure of speech and decipher how that structure is created. By examining
speech, they've ascertained the basic rules of syntax (the way in which
linguistic elements are put together to form phrases) and semantics (the
study of word meanings). Together, syntax and semantics make up the
foundation for a language. Anything that has a language has the ability
to communicate.

The universe itself has a language, and it speaks with the Bioglory of
God:

> The heavens declare the glory of God; and the firmament
> sheweth his handywork. Day unto day uttereth speech, and
> night unto night sheweth knowledge. There is no speech nor
> language, where their voice is not heard. Their line is gone
> out through all the earth, and their words to the end of the
> world. In them hath he set a tabernacle for the sun, Which
> is as a bridegroom coming out of his chamber, and rejoiceth
> as a strong man to run a race.
> — Psalm 19: 1-5

Speech can be natural or artificial, but all speech is developed from
a code. Not surprisingly, scientists have found that DNA follows gram-
matical rules similar to those used in human speech. For example, Fosar
and Bludorf discovered that the structure of DNA not only corresponded
to human speech structures, but it also followed all language patterns and
genetic codes. After all, DNA and genetic code existed long before the
first human uttered the first word. DNA functions as a superconductor;
it is able to store and encode both light and information through prayer!

Open Heaven

Jesus always prayed prior to every kingdom demonstration. The scriptures indicate that He would find a place to be alone with the Father and spend significant time in prayer. During those times, Jesus received strategies and tactics directly from the Father, which He would use throughout the day to overcome evil.

Through prayer, we can receive direct downloads of God-infused knowledge to help us plan for future events, gain insights into mysteries, and execute strategies for success in every area of our lives. "Let us therefore come boldly to the throne of grace that we may obtain mercy and find grace to help in the time of need" (Hebrews 4:16). Prayer opens up this reality; Jesus has given us free access to receive from the Father while we are here on earth. "I am the door. If anyone enters by Me, he will be saved, and will go in and out and find pasture" (John 10:9).

Access to hear, access to see, access to receive from the treasures of heaven and then bring those treasures to earth—this is what an "open heaven" concept means to me. God's Bioglory has everything we need. I can directly download the knowledge I need to my DNA. Like John on the Isle of Patmos in the Book of Revelation, we can hear His Voice and ascend higher.

Releasing the Bioglory of Heaven

God wants His church to receive the blessings of an open heaven and use His voice to initiate change. That is why He imparts the Bioglory of Christ to the church. The church in the Book of Acts was so immersed in the Holy Spirit that supernatural miracles continually manifested around them. Jesus taught the twelve apostles to use prayer-encoded Bioglory and to minister out of that heightened state. They would spend countless hours in prayer to receive Bioglory from God and then share that Life-Light with the world.

Prayer-encoding Biology now defines the ministry of the church on earth just as it defined the ministry of Christ. Paul understood the relationship between prayer and Bioglory, and he used this knowledge to carry out his supernatural ministry. He said, "Therefore, since we have this ministry, as we have received mercy, we do not lose heart" (2 Corinthians

4:1). "Such confidence we have through Christ toward God. Not that of ourselves we are qualified to take credit for anything as coming from us; rather, our qualification comes from God, who has indeed qualified us as ministers of a new covenant, not of letter but of spirit; for the letter brings death, but the Spirit gives life" (2 Corinthians 2:14-16).

Ministry that is guided by authentic prayer is prophetic and miraculous, a demonstration of the glorious riches of heaven. Jesus prayed, "For Yours is the kingdom and the power and the glory forever" (Matthew 6:13).

Prayer-encoding Bioglory is our ministry. We are to obtain the Bioglory of God on earth and display His Life-Light to the world, just as Jesus has released the Life-Light during His ministry. God's will has never changed. His desire is to display His power and Bioglory in the twenty-first century to the same extent as when Christ walked the earth.

David proclaimed, "All your works praise you, LORD; your faithful people extol you. They tell of the glory of your kingdom and speak of your might, so that all people may know of your mighty acts and the glorious splendor of your kingdom. Your kingdom is an everlasting kingdom, and your dominion endures through all generations" (Psalm 145:10-13). Paul told the disciples to "walk worthy of God who calls you into His own kingdom and glory" (1 Thessalonians 2:12). Your prayers are indistinguishable from the kingdom, power, and glory of God.

One day after intercessory prayer, someone asked what I meant when I said that I align my prayer with the lights and sounds of heaven. I explained that the ear has both spiritual hearing and alignment capabilities. When we pray, I truly believe that we have the potential to bring our words into alignment with heaven and release the supernatural benefits of the Word of God here on earth.

Light and sound are just two of the tools that I enjoy using during prayer, especially for warfare and healing prayers. When Christ comes to dwell in a person, He speaks through him and His words take on the expression of natural sound. This is the sound within the sound. We are called to release the supernatural sound of heaven through prayer.

Sound waves are mechanical waves that need a material medium to transmit, such air, liquid, or metal. Without that medium, there can be no sound. As a church, we must make a sound that is completely unique to heaven but can be captured on earth by the worshipping saints. "Blessed are the people who know the joyful sound!" (Psalm 89:15). God has a strategic purpose for His gathering saints: prayer. The sound of our prayers carries the Glory of God and releases it into the very atmosphere of heaven and earth. "He who has an ear, let him hear what the Spirit says to the churches" (Revelation 3:6). So, another way to align the systems of the universe is to listen to the Lord, spread his Bioglory, and share his message across the earth through sound.

Sound and Matter

Science has proven that sound waves can be so powerful that they reconfigure matter. Dr. Hans Jenny was a physician and natural scientist who coined the term "cymatics" to describe the acoustic effects of sound wave phenomena. His discovery explained the natural and supernatural effects that sound waves and vibrations have upon matter. Dr. Jenny further explained that different frequencies of sound waves create specific vibratory patterns in water, salt, and sand particles. These experiments with sound can have extreme implications when we consider that the human body is comprised of 60 percent water and our DNA is like crystalline salt.

Because your body is comprised of spirit and matter, you have the ability to hear sounds in both worlds. Jesus said that the words He speaks are Spirit and Life. When we speak the Word of God, we are speaking Spirit and Life.

The Word is active, living, and sharp; it cuts and heals matter. Natural sound waves can reconfigure matter, but only the supernatural sound waves of the Word can *create* matter. Natural sound waves are not miraculous. However, supernatural sound waves reflect quantum reality.

The Word of God is a different kind of sound. It has power and authority over every dimension, system, and realm. Isaiah 55:10–11 says:

> For as the rain comes down, and the snow from heaven,
> And do not return there, But water the earth, And make

it bring forth and bud, That it may give seed to the sower. And bread to the eater, So shall My word be that goes forth from My mouth; It shall not return to Me void, But it shall accomplish what I please, And it shall prosper *in the thing* for which I sent it.

The Word from the mouth of God is a sound wave within a sound wave, creating patterns within the quantum world. It ultimately becomes a supernatural conduit between heaven and the material world, exhibited through the prayers of the righteous.

In other words, faithful members of the church can pray the Word and create entirely new realities out of nothing. The prayers of the righteous can reveal another level of creativity in performing miracles. "Then the LORD said to me, 'You have seen well, for I am watching over My word to perform it'" (Jeremiah 1:11–19). Clearly, there is a different level of reality in operation when we pray the Word of God. Our God is an extra-dimensional, multifaceted Spirit, and He watches over His Word in every dimension and realm.

The Bible doesn't teach us that things were and are created out of nothing; it teaches us that the material world is made of invisible things from the quantum realm. God formed the material world with His Word. He spoke the Word, and sound waves pulled quantum realities into discernible patterns. These sound waves were hardwired into the universe, and they can be activated through prayer.

The Bible teaches us that the entire universe is sustained and held together by the Word of God (Colossians 1:17). These supernatural sound waves are everywhere. They do not need a medium to travel from one place to another. They are the extra-dimensional qualities of an extra-dimensional Spirit God.

God's people can pray the Word and release supernatural frequencies that will continue to resound over all creation and uphold a level of healing and power that has never been seen or heard in the history of the church.

Hardwired

Second Chronicles contains an awesome promise that supernatural sound waves will be created when the people of God pray:

> If my people, which are called by my name, shall humble themselves, pray and seek my face and turn from their wicked ways, then I will hear from heaven, and will heal their land. Now mine eyes shall be open, and mine ears attend unto the prayer that is made in this place.
> —2 Chronicles 7:14

Dr. David Jeremiah wrote, "At the moment you pray, you connect to the most powerful force in the universe. God has hardwired the universe to work through prayer. It's breathtaking to realize that the all-powerful God intends you to have such a huge part in the work of ushering in His kingdom for all eternity."

The supernatural sound waves of prayer are a big deal. Your voice is the vehicle of expression for the sounds of heaven.

National Gods and the Prince of the Airways

When mankind's DNA was diminished in the Garden of Eden, Satan positioned his chief spirits in the heavenly realm to block the back-and-forth flow of prayers. These principalities are gods, who now rule the airways.

Satan is also known as the prince of the airways (and sound frequencies). His principalities operate as "national gods" under his dominion. These principalities have direct control and rule over lesser spirits to carry out the present agenda of the kingdom of darkness.

Earlier, we discussed God dispersing the nations at Babel and allotting seventy nations to the sky gods, but He kept a bloodline for Himself. This allotment was divided along geographical territories where the fallen angels were worshipped. We also established the connection between the physical heavens and the spiritual heavens in the Bible. Some scriptures link the stars of heaven with angelic beings, referred to them as "holy ones, or "sons of God" (Psalm 89:5-7; Job 1:6).

These spiritual "hosts of heaven" were allotted authority over pagan nations, and they were considered spiritual princes or national gods (Daniel 10). The pagan nations were named after the gods who ruled them. When Daniel had to fight the national god who ruled over Persia, a messenger angel of the Lord came to him twenty-one days after he prayed. The messenger angel was delayed because he had to battle with the Persian national god. The battle took place in the region where this prince or national god had authority; therefore, the spiritual battle was legal and very intense. However, the prince of Persia was no match for the archangel of God.

This scripture also mentions that the prince of Greece will come to fight against the angel of the Lord upon his return to the third heavens:

> "Do not be afraid, Daniel," he said, "for from the first day that you purposed to understand and to humble yourself before your God, your words were heard, and I have come in response to them. But the prince of the Persian kingdom resisted me twenty-one days. Then Michael, one of the chief princes, came to help me, because I was detained there with the king of Persia . . ." Then he said, "Do you know why I have come to you? And now I must return to fight with the prince of Persia; and when I have gone forth, indeed the prince of Greece will come. But I will tell you what is noted in the Scripture of Truth. No one upholds me against these, except Michael your prince."
> —Daniel 10: 1-21

When Daniel prayed to the Lord concerning the nations of Persia and Greece, those principalities went on an all-out attack to hinder the answer to Daniel's prayer.

The book of Daniel gives us an example of how the airways are constantly under attack by these national gods. These gods are still fighting against the messengers of God. So the answers to our prayers may be held up once in a while. However, God always wins. If you can sustain a prayer as Daniel did for twenty-one days, you can win the battle in the airways.

Daniel continued to fast and pray until he heard from God. You, too, must not waver in your devotion to Him.

The Spirit in Bimini

I had a similar experience in 1993, which made me aware of the presence of a national god and the importance of sustaining prayer. I served as one of the lead intercessors at Universal Household of Faith, a ministry located in Freeport Grand Bahama. During this time, a team of intercessors traveled to Bimini, Bahamas, to do some ministry work.

The first day, we all woke up at 3:00 AM for early morning prayer. (I was also fasting.) Everyone gathered outside the hotel on a public dock to pray.

While we were praying and pacing the area, I started to walk closer to the water's edge. In that moment, I looked up and saw a giant woman of native descent standing with one of her feet on the land and the other on the water. She towered high above the land and seemed to possess physical and spiritual properties. Her face was very beautiful. Her long black hair moved with the wind, like mine. She had a huge living snake wrapped around her body, and she was adorned with symbols and markings on her arms and legs.

She was staring down at me as if she was in awe that I was able to see her. Soon I heard the voice of the Holy Spirit: "To take back this territory, you must first bind her up. She has authority on both the land and sea for the Island of Bimini." Soon after, the spiritual being disappeared.

I was only fifteen years old. It was my first time visiting the island, and I had no idea at the time what my short engagement with this native woman meant. All I knew was that this spiritual being was in the unseen world, but she somehow had the capacity to interact with the physical world.

Soon after prayer, we went back to our rooms, but I couldn't sleep because of my encounter with this woman. At daybreak, the team went house-to-house, praying for families; people were healed, set free, and delivered from evil. I left the island, but my engagement with this spiritual power was not over.

As human beings, we tend to move on from one event to another in the physical world quite quickly. Our minds and hearts are often fixed on the temporal more than on the eternal. So I moved on with my life, forgetting my assignment "to take back this territory, and bind up the spirit on the Island of Bimini." Spiritual beings, however, do not move on until their assignment is complete.

It was not until six years later, after I was married with two kids, that this experience came back to me. My marriage had been under intense witchcraft and spiritual warfare attacks. Things were being thrown at our marriage, left and right. Then one day, while I was watching a movie about Native Americans and their ownership of lands in the Caribbean and America, I recalled my encounter in Bimini. I realized that all this stress and strain was emanating from this god. I was given a prayer assignment to bind up a national god. I forgot my assignment, but she never forgot hers—to bring me down as an intercessor by any means necessary.

The islands of the Bahamas were inhabited and governed by natives (Mayans, Caribs, and Arawak Indians) before Christopher Columbus came, and they all died out. The native woman I saw towering over the Island of Bimini was a native national god, and the islands and surrounding seas had been under her jurisdiction for centuries. In 1993, the frequencies and sound waves of our intercessory prayers started a spiritual war with this entity, and because I walked away from my duties, I paid dearly until I finally learned the importance of following through with a prayer assignment.

Completing your prayer assignment will bring glory to the name of the Lord. In John 17:4, Jesus said, "I glorified you on earth, having accomplished the work that you gave me to do. And now, Father, glorify Me in Your presence with the glory I had with You before the world existed.... God is working through the prayers of the righteous to have assignments completed on earth."

To add to our conversation of sound waves, I'd like to mention my late Aunt Geniva. When I was a child, she insisted that TV was not of God and refused to watch it. When I visited her a few years ago, she not only had a TV in her room, but she'd purchased one with a sixty-inch screen. She was watching TBN (Trinity Broadcasting Network, and she told me how wonderful it was to hear the Word of God through television programming.

Many believers felt television was evil in the early days because the gods of this world filled the airwaves with satanic, occult programs. The church should have been leading the television and media industries. We need to be the ones occupying these territories, using technology to take the gospel to the masses, and filling the earth with the knowledge of God's Glory.

The Lord told me a number of years ago that He would bring understanding to his people through technology and science. As a result, I knew that scientists would begin to find evidence of the supernatural, and the church would learn how to fight it through prayer. His truths are now being backed up by new evidence from science and cutting-edge technology.

In Daniel 12:4, God instructed Daniel to "shut up the words, and seal the book until the time of the end. The Lord said many will be seeking as knowledge will increase."

The More We Know

Seventy years ago, advances in DNA genetics couldn't even be imagined. God's intention is to continually increase our knowledge until we are able to use supernatural sound waves to release God's Bioglory in the earth realm. The more knowledge we receive from the Holy Spirit, the greater weight of glory we carry. We have a deeper knowledge of supernatural frequencies locked up inside us, which we "*have yet* to *release.*" The Bible calls these sounds "the song of the Lord" (2 Chronicles 29: 27). We were made in His image and likeness, and He encoded abilities within us before we were in our mothers' wombs.

We learned earlier that the heart has an energy field, and that energy can vibrate in tone with the presence of God. Have you ever noticed that when you pray, there is a heavenly melody? These melodies are the supernatural sound waves and "the songs of the Lord" that have been encoded inside of you.

The writer of the book of Hebrews quotes a conversation between Jesus and the Father in which Jesus says, "I will declare Your name to My brethren; in the midst of the assembly I will sing praise to You" (Hebrews 2: 12). This was the sound of the Lord being released on earth, and He released this new sound through prophetic prayers in the midst of his disciples. When the disciples heard Jesus's prophetic prayers, they cried, "Master teach us how to pray!"

Studying the science behind sounds, tones, and vibrations has opened my eyes to another level of Biblical truth. I understand how Jesus mastered supernatural sound waves, tones, and vibrations to destroy the works of the devil. For example, He spoke the Word, and the man with the withered hand was healed (Luke 6:8). He shouted with a loud voice, "Lazarus come forth," and all 60 trillion decomposing cells in Lazarus's body started to materialize. Jesus's voice vibrated from the cross, "It is finished," and the temple veil was torn. His voice accomplished miracles through science.

The Harmony of His Presence
The woman with the issues in her blood-DNA touched the hem of Jesus's garment, and His healing power was transmitted to her body through vibrational frequencies, restructuring her DNA and bringing harmony to her cells (Matthew 14:36). She was made whole. These miracles were accomplished through a scientific process.

John G. Lake wrote, "Science is the discovery of how God does things." I concur that science does not create any new laws or principles but only uncovers things that God has put in place from the beginning of time. God is the same yesterday, today, and forever.

Jeremiah 33:3 reads, "Call to Me, and I will answer you, and show you great and mighty things, which you do not know." God is waiting to show you great things, but you must first pray, believing that He will reward those who seek His Face. The unbelieving culture of many in the church has directed people away from seeking Him in prayer. Mark 6:5-6 explains that Jesus could not do mighty works in that region because of the unbelief of the people of Gadarenes. Obviously, our unbelief has made

His Word impotent. Let us not limit God with our unbelief. When we walk in unbelief, our minds, bodies, and souls are not in tune with the presence of God.

Our God is a God of science and technology. He created everything from the invisible world. He holds the universe together with His Word. God is a big God, but to us, He will only show Himself in accordance with the authenticity of our prayers. "Therefore I tell you, whatever you ask for in prayer, believe that you have received it, and it will be yours" (Mark 11:24). We as believers can access heaven's technology through our faith-filled prayers.

Earlier, we documented that the Holy Spirit uses spiritual sound waves or vibrational resonances from the Word of God to generate Bioglory Morphic Resonance through electromagnetic frequencies (BM-Fields). Because the body is 60 percent water, DNA converts sound energy into light energy, which is the same as converting sound into light.

Sound can be used to produce light. This phenomenon is called *sonoluminescence*. I believe that when we pray and release the sound waves of heaven, Bioglory Life-Light is increased in the earth realm.

Dismantling Principalities with Sonic Frequencies

Solfeggio frequencies are part of the six-tone scale thought to have been used in sacred music, prayers, and songs to create harmony. When mastered by a musician or prayer intercessor, the notes below can reveal sacred frequencies. For example, the note "MI" is for "Miracles" and vibrates at 528HZ. This frequency is used by genetic engineers and biochemists throughout the world to repair the genetic blueprint of DNA.

- UT – 396 Hz – Liberating Guilt and Fear
- RE – 417 Hz – Undoing Situations and Facilitating Change
- MI – 528 Hz – Transformation and Miracles (DNA Repair)
- FA – 639 Hz – Connecting/Relationships
- SOL – 741 Hz – Awakening Intuition
- LA – 852 Hz – Returning to Spiritual Order

I believe that these six original notes were played by the seven ancient priests to shatter the walls of Jericho (see Joshua 6:16). The destructive forces came from the sound and vibration resonance of the BM-Field.

John Keely, an expert in the study of electromagnetic technologies, wrote in his "Formula of Aqueous Disintegration" that the vibration of "thirds, sixths, and ninths notes were extraordinarily powerful." In fact, he proved that the vibratory antagonistic thirds was "thousands of times" more forceful in separating hydrogen from oxygen in water than heat! In all molecular dissociation or disintegration of both simple and compound elements, whether gas or solid, a stream of vibratory antagonistic thirds, sixths, or ninths on the chord mass will compel progressive division.

In other words, sound can disintegrate molecules.

Now imagine the matrix of the walls of Jericho falling as the seven priests played the sacred tones on the seven trumpets while carrying the Ark of the Covenant. The Ark was well known for its power and energy, which likely amplified the sound waves of the trumpets. Even so, your body is the temple of God; therefore, His power and energy (dunamis) is amplified in you by the Holy Spirit.

The BM-Field increases by one degree every day, according to the Mosaic calendar—just like the Solfeggio increased the BM-Field degree by degree, as they played their instruments and marched around the wall of Jericho.

The dismantling of the wall is symbolic of the dismantling of the national gods over the city of Jericho. To take the city, you must first bind the strong man (giant Nephilim gate keepers).

The Science of the Pied Piper

Most of us are familiar with the tale of the Pied Piper. In the same vein, Lucifer is a master musician who has knowledge of the power of sound and vibration, and he has been controlling the world, manipulating the masses, and restructuring molecules and DNA through sound and vibrational frequencies since the days of Adam.

Some studies have proven that certain modern binaural sounds and beats can be used to digitally create the same physiological effects as heroin, cocaine-crack, and LSD. Some tones have even gone over and above the DNA to make individuals express sexual, violent, or depressive natures. I can see this knowledge being used by Satan more and more frequently as time progresses.

Remember, DNA is extremely sensitive to sound and light. Even a slight shift in musical notes can trigger a signal throughout your body. Likewise, modern binaural sounds and beats that have been crafted by the enemy will encode negative information into your DNA. Most of us are completely unaware of how these diabolical sounds and frequencies affect our moods throughout the day. Because of the heightened sensitivity of your DNA, every binaural sound and beat will stir your DNA into some form of biological response. Be very careful because these modern diabolical binaural sounds and beats can be co-creating your life.

The use of binaural sounds and beats (frequencies) to reprogram the DNA is nothing new. This is an ancient science, and it was not a coincidence that a high-frequency program was developed by King Nebuchadnezzar to keep the people in captivity and sway them into idol worship. In Babylonian captivity, a tone or sound frequency was what kept them in bondage:

> People of all nations, races, and languages! You will hear the sound of the trumpets, followed by the playing of oboes, lyres, zithers, and harps; and then all the other instruments will join in. As soon as the music starts, you are to bow down and worship the gold statue that king Nebuchadnezzar has set up... And so, as soon as they heard the sound of the instruments, the people of all the nations, races, and languages bowed down and worshipped the gold statue.
> —Daniel 3:1-7

I believe that King Nebuchadnezzar used the Pied Piper strategy to sway the nation into idol worship. Today the church cannot be ignorant concerning the schemes of the enemy. Through prayer, we must send out supernatural high frequency sound waves to combat the attacks to our airways (2 Corinthians 4:4).

Blow the Trumpet in Zion

In Numbers 10:9, the Lord tells us to blow the trumpet when the battle gets hot, and He will remember and help. When Gideon sounded the trumpet, he defeated 120,000 Midianites with only 300 men. This story leads us to the conclusion that the supernatural sound waves of prayer are the most powerful technological weapons the church has ever known.

In Jeremiah 4:19 the prophet exclaims, "Oh, my anguish, my anguish! I writhe in pain. Oh, the agony of my heart! My heart pounds within me, I cannot keep silent. For I have heard the sound of the trumpet; I have heard the battle cry."

As you actively engage in prayer, God will reveal His purposes through sound—not just His Voice alone, but also through other sonic frequencies that will dismantle national gods and shake evil out of their illegal places on Earth.

David said unto God, "Therefore let all the godly pray to You while You may be found. Surely when great waters rise, they will not come near. You are my hiding place. You protect me from trouble; You surround me with songs of deliverance. Selah" (Psalm 32: 7).

These "songs of deliverance" were sacred and prophetic. They were high-frequency sounds released from heaven while David prayed. In this scripture, the term *deliverance* describes the forgiveness of sin, redemption from eternal death, release from spiritual bondage, recovery of physical health, and rescue from difficult situations. These high-frequency sounds, which are still used today, can set captives free. They are so powerful, they can result in deliverance from demonic bondage.

The church is challenged to capture these sounds and release them through prayer and worship music. When this release happens, the very supernatural sound of heaven penetrates the natural sound surrounding the earth, and the atmosphere of heaven causes everything to transform.

Again, high-frequency sound will set captives free: "And at midnight Paul and Silas prayed, and sang praises unto God: and the prisoners heard the sound. And suddenly there was a great earthquake, so that the foundations of the prison were shaken: and immediately all the doors were opened, and every one's bands were loosed" (Act 16:26-27).

Peter had a similar experience with life-saving sound:

> So Peter was kept in prison, but the church was fervently praying to God for him. On the night before Herod was to bring him to trial, Peter was sleeping between two soldiers, bound with two chains, with sentries standing guard at the entrance to the prison. Suddenly an angel of the Lord appeared and a light shone in the cell. He tapped Peter on the side and woke him up, saying, "Get up quickly." And the chains fell off his wrists. Get dressed and put on your sandals," said the angel. Peter did so, and the angel told him, "Wrap your cloak around you and follow me." So Peter followed him out, but he was unaware that what the angel was doing was real. He only thought he was seeing a vision. They passed the first and second guards and came to the iron gate leading to the city, which opened for them by itself. When they had gone outside and walked the length of one block, the angel suddenly left him…. And when he had realized this, he went to the house of Mary the mother of John, also called Mark, where many people had gathered together and were praying.
>
> —Acts 12:5-12

As the church delves into this assignment, we must explore different high-frequency prayer and praise strategies, which will be revealed by the Holy Spirit.

Incorporating the Sound of Music into Prayer

In 1 Chronicles, we see David using prayer and music to create a new sacred frequency within his kingdom. He trained the priests, commanders, gatekeepers, and musicians to prophesy with harps, lures, stringed instruments, and cymbals:

> And David spake to the chief of the Levites to appoint their brethren *to be* the singers with instruments of musick,

psalteries and harps and cymbals, sounding, by lifting up the voice with joy...And Chenaniah, chief of the Levites, *was* for song: he instructed about the song, because he *was* skillful.

—1 Chronicles 15:16-22, 27

All these *were* under the direction of their father for the music *in* the house of the LORD, with cymbals, stringed instruments, and harps, for the service of the house of God. Asaph, Jeduthun, and Heman *were* under the authority of the king.

—1 Chronicles 25:6

David placed a greater emphasis on practicing prayer and music than any other figure in the Bible. He sought to capture the sacred, high frequencies of heaven here on earth. As a chief master musician, intercessor, and worshipper, David studied natural and supernatural sounds. The instruments used by David were called "the musical instruments of God" (1 Chronicles 16:42). The instruments in the temple were purposely built by David for communing with God through prayer and worship (2 Chronicles 7:6).

Capturing the Frequencies

In every generation, some are called to capture and express the sacred frequencies of heaven on earth. As a member of the church, you have been called to do His work. Sound precedes light, and if we want to see a Bioglory explosion, we must release the supernatural sound of God.

Ezekiel captured the sacred sound of God in His temple. While in a vision, he beheld the glory of the God of Israel, which came from the east. His voice was like the sound of many waters; and the earth shone with His glory (Ezekiel 43:2). Ezekiel used that frequency when he prophesied or ministered unto his people. For example, when he prophesied to the dry bone, Ezekiel heard the sound of the bones connecting, and an army came together in the valley (Ezekiel 37:7). The high frequency sounds of heaven can materialize and restructure molecules, DNA, and matter to do God's bidding. Ezekiel mastered the art of capturing the sacred sounds of heaven on earth, and we are called to do the same.

In like manner, the disciples captured the sacred sounds of heaven. While they were praying, the Holy Spirit came, creating an unexpected sound, and then fire (light) appeared. "Suddenly a sound like a mighty rushing wind came from heaven, and it filled the whole house where they were sitting . . . tongues like as of fire, and it sat upon each of them" (Acts 2:2). This prophetic encounter started with an auditory signal that was followed by visual revelation.

While John was on Patmos, he was caught up in the Spirit and revealed: "I was in the Spirit on the Lord's day, and heard behind me a great voice, as of a trumpet . . . And I turned to see the voice that spake with me. And being turned, I saw seven golden candlesticks; And in the midst of the seven candlesticks one like unto the Son of man, clothed with a garment down to the foot, and girt about the paps with a golden girdle." (Revelation 1:10-13). He heard a sound, then He saw seven golden candlesticks and the Son of man. John went on to recount how he heard another high frequency sound from heaven that was "like the sound of many waters and like the sound of a great thunder. I heard the sound of harpists playing their harps" (Revelation 14:2).

When the church is led by the Spirit into prayer, mighty, prophetic miracles can occur. But completing smaller tasks are also important. Sometimes, the Holy Spirit will direct the church to complete a simple action as a declaration of their commitment to heaven. In Ezekiel 21:14, the Lord said to Ezekiel, "You therefore, son of man, prophesy and clap your hands together." A few verses down the Lord says, "I will also clap My hands together, and I will appease My wrath; I, the Lord, have spoken" (Ezekiel 21: 17). These kinds of actions are symbolic. They show that when believers are engaged in an act on earth, the Lord is engaged in the very same act in heaven, but with more significant results. Your prophetic declarations are multifaceted and multidimensional, so don't undermine them.

For example, when the church releases a mighty shout on earth, God Himself will release a might shout in heaven. We see this truth in Isaiah:

Sing to the Lord a new song, and His praise from the ends
of the earth... Let the wilderness and its cities lift up their
voice ... Let the inhabitants of Sela sing, let them shout from
the top of the mountains. Let them give glory to the Lord,
and declare His praise in the coastlands. The Lord shall go
forth as a mighty man, he shall stir up jealousy like a man
of war: he shall cry, yea, roar; he shall prevail against his
enemies. I have long time holden my peace; I have been still,
and refrained myself: now will I cry like a travailing woman;
I will destroy and devour at once.
—Isaiah 42:10-14

If you pray authentically, the Lord will back you up when you are
engaging in spiritual warfare. He has promised to release a sound from
heaven that will dismantle every enemy, sickness, and disease in your life.
In another verse, we see the Lord exhorting His people to worship with
the promise that He will conquer all of Israel's enemies (Isaiah 30:29-32).
In other words, God has your back.

Have you ever witnessed an entire congregation of believers in a
warfare chant? From the young to the elderly, all are engaged. The musicians are playing and the sound of the instruments are in sync with the
prayers and praise of the people. It is within this atmosphere that the
highest frequencies are strategically created to dismantle the enemies of
God:

From the lips of children and infants you have ordained
praise because of your enemies, to silence the foe and the
avenger.
—Psalm 8:2

When I remember these things, I pour out my soul in me:
for I had gone with the multitude, I went with them to the
house of God, with the voice of joy and praise, with a multitude.
—Psalm 42:4

Your prayers and praises are prophetic weapons that can level cosmic mountains, break down gates of bronze, cut through bars of iron, shatter glass ceilings, and shift strongholds in the natural world.

> Let the saints be joyful in glory; Let them sing aloud on their beds. Let the high praises of God be in their mouth, And a two-edged sword in their hand, To execute vengeance on the nations, And punishments on the peoples; To bind their kings with chains, And their nobles with fetters of iron; To execute on them the written judgment—This honor have all His saints. Praise the Lord.
> —Psalm 149:5-9

The Spirit in Warfare

One of the greatest revelations in the Bible is that the same Spirit that rose Christ from the dead lives within you. The Holy Spirit within you can create the sound of earth and heaven, which is extremely powerful. As we already examined in Joshua 6, the sounds of heaven and earth combined to make a sound frequency so powerful that it disintegrated the materials that held the wall together—every atom and sub-atomic particle was disbanded. "If anyone speaks, he should do it as one speaking the very words of God" (1 Peter 4:11).

Christ in you is heaven's New Testament technology. The Life-Light of Christ is stored and duplicated within His people, who live on the earth. We are the living body of Christ. We are Christ in the flesh—many members but one body. Those who have received Christ as Lord become His hands and feet on earth. They also become His mouth, and they speak with His two-edged sword, which is His Word.

Christ speaks prophetically through His church. The Word of God heals, delivers, and sets the captives free. Your DNA and cells were uniquely designed to align systems and carry God's Bioglory to earth. As you yield your body to God, He will use your vocal chords to demonstrate His majesty, wisdom, and knowledge through prayer.

CHAPTER 12
THE MINISTRY OF BIOGLORY

For the Lord will arise over you, and His glory will be seen upon you.
—Isaiah 61:1-3

Radical prayer ministry begins with the revelation that God wants His power and glory to be seen upon the earth. Paul said, "He called you by the gospel for the obtaining of the glory of our Lord Jesus Christ" (2 Thessalonians 2:14). "The riches of His glory" is the true inheritance of the saints (Ephesians 1:18). The New Testament Covenant is a ministry of Bioglory. "Our sufficiency is from God, who has made us sufficient to be ministers of a new covenant, not of the letter but of the Spirit. For the letter kills, but the Spirit gives life" (2 Corinthians 3:4-6).

Now if the ministry of death, carved in letters on stone, came with such glory that the Israelites could not gaze at Moses's face, the ministry of the Spirit, which hails from pure righteousness, must far exceed it in glory. Indeed, in this case, the ministry of death was carved onto stone tablets, but the ministry of the Spirit was carved into human crystalline DNA. The glory of the New Testament surpassed the glory of all that came before it. Our DNA has *permanent* glory, glory that has been sealed by the Holy Spirit, glory that will last for time and all eternity (2 Corinthians 3:5-11).

Jesus prayed that His disciples would be with Him and behold His glory (John 17:24). Therefore, we find peace with God through our Lord Jesus Christ, through whom we also have access to His grace and glory (Romans 5:1-5). All we need to do is spend time with the Father and receive His divine Bioglory, which demolishes the stronghold of the flesh and releases the Life-Light of Christ within us.

The disciples of Jesus went from one profound Bioglory encounter to another over a period of three years. They constantly witnessed miracles.

"We beheld His glory; the glory of the One and Only who came from the Father" (John 1:14).

Because the disciples followed Jesus, they ultimately reproduced His Bioglory. "A disciple is not above his teacher, nor a servant above his master. It is enough for the disciple to be like his teacher, and the servant like his master" (Matthew 10:24-25). The early apostles were "conformed" to the exact image of Christ. They were transformed into the same image from glory to glory.

Today, we are living in a time when the Bioglory of God is beginning to become manifest in a way that has not been seen since the church in the Book of Acts. As people are gathering to pray and seek the face of God, they are beginning to experience the radiant realm of heaven touching earth. As we draw nearer to God in our hearts, we inevitably cross the threshold. "Therefore, brothers, since we have confidence to enter the Most Holy Place by the blood of Jesus . . . Let us draw near to God" (Hebrews 10:19, 22).

Intimacy with the Father

Prayer is an invitation to enter the most holy place, which is filled with the Shekinah Glory of God. The Bioglory of God is not just a powerful theological concept; it is a powerful spiritual reality that God desires His people to experience. As we are filled with His Bioglory on earth, we literally become one with God, just as Jesus and the Father are one.

Prayer-encoding Bioglory is the catalyst that catapults us into a deep intimate oneness with God. This oneness cannot be achieved outside of prayer. Every Shekinah Glory encounter begins with seeking His face in prayer. "When You said, 'Seek My face,' My heart said to You, 'Your face, Lord, I will seek'" (Psalm 27:8). There is no secret formula or hidden technique behind prayer. You just need to seek the Lord with a heart of worship; that is the *only* way we will experience God.

The Bioglory Life-Light of God flows from Him to those who have hearts that long for Him. As you seek Him, you will build an intimate relationship with Him. "This is the eternal life, that they may know You, the only true God, and Jesus Christ whom You have sent" (John 17:3).

As you pray, you will pass beyond "the veil," crossing the threshold into the most holy place where intense Bioglory emanates from God. In this secret place, the supernatural power of the Holy Spirit will overshadow you and everything around you. God's Bioglory will radiate inside of you so much that you will have a glow that pierces through your temporal circumstances (Exodus 34:35, Acts 2).

When matter interacts with Biolgory, an atmosphere for the supernatural is created. People are healed, the anointed destroy the yoke, and multitudes receive the gospel.

The Breath of God

The Father achieved the impossible through Christ. Jesus exercised perfect authority over nature at a subatomic level. Through the power of the Holy Spirit, Jesus's voice was the very voice of God on earth. It was the same creative voice that conceived the impossible in Genesis 1. All things came into existence by His command.

Jesus said, "The words that I speak to you, I do not speak on My own authority; but the Father who dwells in Me does the works" (John 14: 10). The very breath of God orchestrates the works of the kingdom, raising the dead, casting out demons, healing the sick, and cleansing the lepers.

John the Baptist said of Jesus, "He whom God has sent speaks the words of God, for God does not give the Spirit by measure" (John 3: 34). All that Jesus said and did on earth He did as a man, not as God (John 1:14). He prayed and encoded Bioglory as a man. He healed the sick and walked on water as a man. Jesus was filled with the Holy Spirit. He demonstrated what a human life filled completely with the ministry of Bioglory could achieve. "Do you not believe that I am in the Father and the Father is in Me? The words I say to you, I do not speak on My own. Instead, it is the Father dwelling in Me, carrying out His work" (John 14:10).

Life and Breath in Motion

Jesus said, "I have not spoken on My own; but the Father who sent Me, gave Me a command, what I should say and what I should speak.

Therefore, whatever I speak, just as the Father has told Me, so I speak" (John 12: 49-50). In addition, Jesus told His disciples, "At that time you will be given what to say, for it will not be you speaking, but the Spirit of your Father speaking through you" (Matthew 10: 19-20). These scriptures show that when you act in His name, you will open your mouth and the very breath of God will flow from you.

In the original version of the Torah, the word *breath* is "voice." His breath, His voice, will flow from you. The Hebrew word for God's voice or breath is Ruach, which can also be translated to mean "life and breath in motion." The Holy Spirit is the breath of God, which disperses His intentions. Therefore, the breath of God is the Holy Spirit in motion. "And the Spirit of God was moving over the surface of the waters . . . hovering over the chaos [creating a Bioglory M-Field] to receive the Word. Then the LORD said, 'Let there be light.'" (Genesis 1:2). The Ruach of the Almighty is also the spirit of wisdom, counsel, might, and understanding given to a person.

The Ruach of God gives life to all creation. We could say that God's Ruach has created every other (non-divine) ruach that exists. All living creatures owe the breath of life to the creative spirit of God. Moses stated this truth explicitly when he said: "God . . . gives breath [ruach] to all living things" (Numbers 27:16). Job understood this truth as well. He said, "As long as I have life within me, the breath [ruach] of God is in my nostrils" (Job 27:3). Later, Elihu tells Job, "The Spirit of God has made me; the breath of the Almighty gives me life" (Job 33:4).

Ruach Encodes Resurrection Power

Jesus could open His mouth and instantly command supernatural healing. He could even resurrect Lazarus, whose cells had been decomposing for days. When Jesus came, Martha said, "Lord, by this time there is a stench, for he has been dead four days!" (John 11:17, 39).

There was no spirit or life in Lazarus's body, but there is spirit and life encoded in the breath of the Almighty. Jesus prayed to the Father and then He called out in a loud voice, "Lazarus, come out!"

Breath encoded with spirit and life went out to accomplish the works of the kingdom. "And the dead man came out, his hands and feet wrapped with strips of linen, and a cloth around his face. Jesus said to them, 'Take off the grave clothes and let him go'" (John 11:43-44).

Again, we see the Holy Spirit in motion when God's voice commanded everything to come into perfect divine order within Lazarus's body. This example shows the extreme power that the breath of God has over the BM-Fields in the quantum world.

Every time the breath of the Father is in motion, the quantum fields are totally re-created or restructured to come into perfect divine order according to His will. You also have the breath of God within you. "If my words abide in you, ask whatever you wish, and it will be done for you" (John 15:7). The words of Jesus have already been given to us. The Holy Scriptures, formed out of Jesus's supernatural teachings, contain a mountain of treasures.

At the center of the apostles' spiritual training was the power of prayer. The Lord's breath was to become their prayer weapon against all the works of the enemy, from the micro to the macro components in the universe. Paul called the Word, the sword of the Spirit (Ephesians 6:16,17).

Supernatural Training

Jesus taught the disciples how to create and re-create realities in the quantum world by praying faith-filled words. He said, "Have faith in God. For assuredly, I say to you, whoever says to this mountain, 'Be removed and be cast into the sea,' and does not doubt in his heart, but believes that those things he says will be done, he will have whatever he says" (Mark 11:22-23).

You can use the breath of the Father to create a new reality in the quantum world by speaking to the mountains that need to be removed. Jesus said, "Whoever says to this mountain . . ." He didn't teach His disciples to pray to God for the mountain to be removed. He taught them to believe in their hearts and command that mountain to get up and move. He taught them to speak directly to the mountain and create a new reality. Whatever they said would come to pass if they only believed in their words. The key in this model of Bioglory modality training was to use the breath of the Father.

Jesus used this technique to minister healing and deliverance from the third heaven to the earth realms. He didn't pray words to the Father for the mountains of sickness and satanic bondage to be removed. He believed in what the Father said, and He spoke those words directly to the situation at hand and things changed. Jesus was filled with the Holy Spirit, speaking and praying only with the breath of His Father. When He opened His mouth, the material universe around Him obeyed.

Jesus used this same method whether He was commanding a demon to leave or a disease to be healed or a dead body to come to life. No matter the problem, He always used a Bioglory modality and spoke directly to the problem. He even spoke to the four winds and the waves and commanded them to be still.

> And the same day, when the even was come, he saith unto them, Let us pass over unto the other side. And when they had sent away the multitude, they took him even as he was in the ship. And there were also with him other little ships. And there arose a great storm of wind, and the waves beat into the ship, so that it was now full. And he was in the hinder part of the ship, asleep on a pillow: and they awake him, and say unto him, Master, carest thou not that we perish? And he arose, and rebuked the wind, and said unto the sea, Peace, be still. And the wind ceased, and there was a great calm. And he said unto them, Why are ye so fearful? how is it that ye have no faith? And they feared exceedingly, and said one to another, What manner of man is this, that even the wind and the sea obey him?
> —Mark 4:35-41

In this scripture, we see that Jesus rebuked his disciples for not applying what he taught them. He expected them to speak directly to the quantum world themselves rather than asking Him to do it for them. Throughout the scriptures, we see that Jesus modeled the ministry of Bioglory to the disciples and expected them to adopt the very approach that He Himself used.

The primary focus of this ministry was praying and ministering with the breath of the Father. The disciples were trained to exercise dominion over the kingdoms of this world.

Jesus spoke with power, and the people were continually amazed at the level of authority He exercised wherever He went. "They were astonished at His teaching, for His words possess power and authority. The high priests and the scribes with the elders exclaimed, 'What authority and power this man's words possess! Even evil spirits obey Him and flee at his command!'" (Luke 4:32, 36).

Jesus was serious about accomplishing the will of the Father so He lived and breathed these words: "On earth as it is in heaven!" In all His works, He skillfully used the keys of the kingdom and then He passed this power and authority on to the disciples by giving them the very same set of keys.

Christ gave His disciples the keys that empowered them to heal the sick, raise the dead, and cast out demons. Jesus only gave the keys of the kingdom to His disciples, to those who were directly connected to Him. The disciples had authority to bind and loose because they had a relationship with Jesus and operated under His authority.

This authority and power is delegated through Christ's disciples. In other words, this kind of authority comes through a representative. For example, behind every soldier is the power of the general. Behind the power of the general is the entire army. Power must be delegated through various levels of authority. Jesus, the Son of God, gave the keys of His kingdom to those who have been reborn. We are backed by the authority of the kingdom, which is backed by God Himself.

No Geographical Limits to Bioglory Modality

In Matthew 8:5-13, the centurion said, "I am a man under authority, having soldiers under me." In essence, the centurion was saying, "Jesus, I recognize that my faith is in your authority. When you speak, there is authority behind your words. The Word of God, the angelic hosts of heaven, and all of the power of the kingdom are behind your words. I understand your level of authority, and when you speak, sickness must obey; it must become your servant and follow your instruction." This story reveals that the centurion understood the principle of authority and that there are no geographical limits to Bioglory modality.

The word *centurion* is based off of the Latin root word *cent,* which means "one hundred." So, this designation indicates that the centurion had authority over one hundred soldiers. However, this man had already recognized the authority that Jesus had over the local and nonlocal realms. He understood the authority through which Jesus operated. He said, "For I am also a man set under authority, having under me soldiers, and I say unto one, Go, and he goes; to another, Come, and he comes; and to my servant, Do this, and he does it."

Then Jesus said to him, "I will come and heal him?"

The centurion replied, "Lord, I do not deserve to have you come under my roof. But just say the word, and my servant will be healed. After hearing this, Jesus said, 'This is great faith!'"

We do not often connect authority to faith, but Jesus said, "If you understand and operate in kingdom authority, that is great faith."

The centurion's faith was grounded in his clear understanding of authority and Bioglory modality. He recognized that although Jesus walked in spiritual authority, and he walked in temporal authority, both worlds still operated from the same principles of authority.

The power we display comes from God, and we have authority over every manifested work of Satan in the earth realm. Because the centurion understood authority, Jesus commended him for having "great faith." The story of the centurion serves as a reminder that we must understand and submit to the authority of Christ in order to receive answers to our prayers and minister to our communities successfully. Jesus said, "If you abide in Me, and My words abide in you, you will ask whatever you desire, and it shall be done for you" (John 15:7).

Jesus let His disciples in on a powerful spiritual secret: they could gain His authority through prayer. But this secret leads us to an even more thrilling prophetic insight into the nature of His mystical relationship with God. "I tell you the truth, the Son can do nothing by Himself; He can do only what He sees His Father doing, because whatever the Father does the Son also does. For the Father loves the Son, and shows Him all things that He Himself does" (John 5:19-20).

This scripture reveals the specific nature of Jesus's relationship with His Father, and this unique, mystical relationship outlines a pattern for how we can release the ministry of Bioglory in a fruitful way.

Jesus said, "I and My Father are one" (John 10: 30). This oneness shows the unparalleled intimacy Jesus had with the Father, a oneness that Jesus revealed to His disciples. The fruitfulness of any ministry is proportionate to the level of connection between the kingdom and the church. The more connected you are with God, the more successful you will be when completing His ministry. Jesus perfectly executed the Father's will on earth because His connection with the Father was at the highest level. Jesus prayed without ceasing and walked in fellowship with the Father.

Jesus said, "If you had known Me, you would have known My Father also; and from now on you know Him and have seen Him" (John 14:7).

Absolutely nothing could ever come between the Father and the Son. This deep oneness of spiritual connection and intimacy was the solid foundation for Jesus's ministry of Bioglory. His ministry was extremely fruitful because he was one with God.

On Earth, Jesus lived as a man who completely depended upon His faith in the Holy Spirit and the Father. We often think He was exempt from certain edicts, like praying and walking in faith, just because He was the Son of God. But he was not exempt. He had to pray daily and walk in faith so that he could lead by example, demonstrating to His followers that they too must pray and have complete faith in God.

Engaging and Dismantling Powers

The scriptures make it clear that we have won the ultimate victory through Christ Jesus. Through Him, we are redeemed. Yet, as long as we are on Earth, the battle for our souls will continue. Therefore, we must be aware of who Satan is and how his kingdom operates.

Knowing the enemy's name is not as important as understanding his spiritual makeup. His spiritual makeup will provide information on how to stand in victory against him. As we previously studied, before God created the heaven and the earth, He created angels:

Then the Lord answered Job out of the whirlwind and said: "Who is this that darkens counsel by words without knowledge? Dress for action like a man; I will question you, and you make it known to me. Where were you when I laid the foundation of the earth? Tell me, if you have under-standing. Who determined its measurements—surely you know! Or who stretched the line upon it? On what were its bases sunk, or who laid its cornerstone, when the morning stars sang together and all the sons of God shouted for joy?"
—Job 38:1-7

Both the morning stars and sons of God in this passage refer to angels. In the beginning of time, there was no need for a priesthood because there was no sin. A priesthood was established as a mediator between God and men, to make concessions for sin through prayers and sacrifices (1 Timothy 2:5).

In the beginning, the sons of God ascended into the Holy Moun-tain to worship God (Isaiah 14, Ezekiel 28). The sons of God were led by the morning star, Lucifer. The name Lucifer is translated from the Hebrew word "helel," which means brightness. There are references to the "morning star" or "star of the morning" or "bright star" in Isaiah 14:12-14 as well.

Lucifer's genetic makeup suggests that he was created to usher angels and other created beings into the presence of God. He has pipes that represent sounds and frequencies, which are designed to assemble the multitudes. Lucifer has power over the airways, and he can influence and manipulate sound frequencies. In fact, he is known as "the prince of the power of the air." He was a genetic light bearer in the Holy Mountain in Heaven. As we previously said, sound can be used to created light. Lucifer emits light by using sound frequencies. Consider the following scripture:

Thou hast been in Eden the garden of God; every precious stone was thy covering"
—Ezekiel 28:13

Lucifer was adorned with a breastplate, and he walked amidst the "stones of fire" in Eden. The Bible reveals that he was fitted with nine of the twelve breastplate stones: sardius, topaz, diamond, beryl, onyx, jasper, sapphire, emerald, and carbuncle.

We documented that the Levitical high priest breastplates were fitted with twelve crystal stones (Ezekiel 28:13; Exodus 28:17-20). We also proved that the DNA in the breastplate stones shared the same qualities as crystalline stones. The number *twelve* often indicates governmental perfection or complete divine governance. We can conclude that the breastplate of twelve stones symbolized the complete government of God over the twelve tribes of Israel. The breastplate served as a medium through which God provided direction to His people. The twelve names engraved upon the stones of the breastplate represented the DNA genetic blueprint of God.

Lucifer's Genetic Limitations
The breastplate has prophetic information embedded within each stone. God created Lucifer with the ability to understand nine out of the twelve prophesies, which means, God placed genetic limitations on Lucifer. Here are the three crystal stones He did not include in Lucifer's breastplate:

- *Opal* – Science has discovered that out of all sound waves, the frequency of love resonates at the highest level. The opal stone within the breastplate represents God's love for His creation. His love was so pure that He sent His Son to die for all mankind. The opal breastplate stone connects the heart, mind, and soul to the spiraling reality of God's love. On a related note, the Life-Light of God is literally wrapped around your DNA in a spiral hologram. Lucifer will never understand the love that God has for humanity or why this love came in the form of a Son who died for our sins, restoring us back to the Father. Lucifer never had this stone within his breastplate; therefore, we will always defeat him with the love of Christ. Love is the more excellent way (1 Corinthians 12).

- *Agate* – Agate is a hexagonal crystal. Remember, the human body is a natural resonator designed to exhibit the Bioglory of God. As a member of the royal priesthood, our cells are Bioglory carriers. Lucifer doesn't have the agate stone in his breastplate; therefore, he will never fully understand how God will transform us into the exact image of Christ. He does not know the mysteries of the kingdom or the ministry of Bioglory (Matthew 13).

- *Amethyst* – Amethyst also translates as "dream stone," which relates to Jacob's DNA ladder and the unmixed new wine of the Holy Spirit within the house of God. Under the Mosaic law, Moses turned water into blood, an act that represented the judgment. Under the principle of grace, Jesus turned water into wine, an act that represented the outpouring of the Holy Spirit and the fullness of joy. This outpouring was part of Melchizedek's blessing on Abraham, and 2,000 years later, Jesus used it in the Passover to represent the blood of the new covenant. In addition, Psalm 75 uses a cup of wine as a symbol of God's wrath. Revelation 18 depicts a similar cup in the hand of Babylon, which is a symbol of how Nimrod and the Babylonians tried to profane the image and likeness of God by altering human DNA. But these kinds of attempts will never succeed because God's followers continue to do His work. Again, Lucifer has no knowledge of this prophetic implication because he lacks this stone. He will never understand how the church will overcome the world through prayer, even as Christ overcame the world through His sacrifice.

Lucifer and the rulers of this world are defeated because of their lack of understanding of these prophetic mysteries. Lucifer has genetic limitations, and this is the advantage that the Lord has given the church over the kingdom of darkness. In 1 Corinthians 2:9, it says, "None of the rulers of this age understood it. For if they had, they would not have crucified the Lord of glory." This scripture is referring to spiritual rulers—what Psalm 82 calls *gods*.

The three additional stones were only worn by the Levitical high priest because they held the prophetic knowledge for God's plan of redemption. They relayed how the Messiah would redeem the world back to God through his DNA (blood). Satan does not have these three stones, and therefore, he does not have knowledge of how our salvation will take place. Similarly, the fallen rulers or gods have no knowledge of this hidden mystery.

As you can see, these three stones play an important role in the restoration of the Father's Bioglory within human DNA. The genetic language of His Life-Light will transform us into the image of Christ, just like it changed Paul and Jacob. For it states in the scriptures: "But you are a chosen people, a royal priesthood, a holy nation, a people for God's own possession, to proclaim the virtues of Him who called you out of darkness into His marvelous light" (1 Peter 2:9).

The Father hides the mysteries of the kingdom in science. If we want to talk about engaging and dismantling powers, we must understand the mysteries of the kingdom.

Your knowledge of God's Glory is a weapon against Lucifer and the rulers of this world. God gave a prophetic promise: "For the earth will be filled with the knowledge of the glory of the Lord as the waters cover the sea" (Habakkuk 2:14). Why is it important for you to know how His glory will cover the earth as waters covers the sea? Paul gives us the answer in 2 Corinthians 4:6, "For God, who said, 'Let light shine out of darkness,' made his light shine in our hearts to give us the light of the knowledge of God's glory displayed in the face of Christ." Through our DNA, we can experience the glory of the Lord.

The Knowledge of Bioglory

DNA is the genetic code for universal language. Information is encoded into genetic material for all living organisms—from simple bacteria to animals to humans—which means that all of His creations can encode the genetic language of Bioglory through DNA. Our understanding that

God will use DNA to restore all His creation is of vital importance. Now, we may use all that we've learned about BM-Fields and the holy sounds and frequencies of prayer to affect His Will.

As Christ's followers today, we are the royal priesthood wearing the twelve stones of His redemptive power. I truly believe that our non-coding DNA holds the very switches that the Holy Spirit will use to "turn on" the limitless power and glory of God within us.

Jesus said, "I will give you the keys to the realms of heaven" (Matthew 16:19). Jesus gave His church the keys to the kingdom of heaven, and to my understanding, these keys are genetic codes that will unlock realms of power and glory within us so that we, in return, can unlock these realms of power and glory in the earth realm. The human body has so many switches that science has yet to discover them all.

Fortunately, God does not seek to withhold the knowledge of His glory, which unlocks these genetic codes within us. We only have to pray to seek the truth. But religion and tradition will take away the knowledge of God's glory, thus keeping many in ignorance to this truth. In one translation of Hosea, God says:

> My people are destroyed from lack of knowledge. Because *you have rejected knowledge*, I also reject you *as my priests* because you have ignored the law of your God, I also will ignore your children.
> —Hosea 4:6

We often skip the main subject in this scripture: our lack of knowledge is a serious matter in the church today. When we reject knowledge, God rejects us as priests. Those who are rejected as priests will not wear the breastplate with the twelve stones, and they will not have the power of Christ to overcome Lucifer. And not only will they be rejected but so will their offspring.

The Generational Gap

There is a generational gap of prophetic knowledge and wisdom that God is seeking to restore to the church in this age. This knowledge is the lost art of prayer-encoding Bioglory. The discipline of spending countless hours with God has been neglected in the church. We must restore the prayer altars, and in doing so, we will receive the "river of fire" from God. As we pray, a Bioglory revival will break out in our lands (Isaiah 58:12).

Jesus went to war because the religious spirit of His day sought to rob God's people of an experience with His glory. Jesus said, "Woe to you experts in the doctrines of the law, because you have taken away the key of knowledge." He also said, "You yourself have not entered [into this experience], and you have hindered those who were entering" (Luke 11:52).

Matthew wrote, "Woe to you, you hypocrites, scribes, and pharisees! You shut up the Realm of heaven in men's faces; you neither enter yourselves, nor will you let those enter who are on the point of entering!" (Matthew 23:13).

So, let me say this again: the keys to the realm of heaven are genetic codes that will unlock realms of power and glory within us. We need the knowledge of His glory. Religion has removed the knowledge of God's glory. To regain this knowledge, we must turn to God in prayer.

Puritan Thomas Manton wrote, "O how little do we aim at the glory of God and regard it in our prayers." I believe that most people don't aim for the glory of God in prayer because they have no knowledge of what the glory is and how it benefits their lives.

Paul prayed for the church to have this knowledge:

> I have not stopped giving thanks for you, remembering you in my prayers in order that the God of our Lord Jesus Christ, the glorious Father, may give you a spirit of wisdom and revelation in your knowledge of Him. I pray also that the eyes of your heart may be enlightened, so that you may know the hope of His calling, the riches of His glorious inheritance in the saints and the surpassing greatness of His power to us who believe. These are in accordance with the working of His mighty strength, which He exerted in Christ when He raised Him from the dead and seated Him at His right hand in the heavenly realms far above all rule and authority, power and dominion, and every name that

is named, not only in this age, but also in the one to come. And God put everything under His feet and made Him head over everything for the church, which is His body, the fullness of Him who fills all in all.

—Ephesians 1:17-23

Why would we want to walk in ignorance when we have been given the knowledge of His glory? Why follow hollow traditions instead of supplicating the Lord in prayer? Why ignore science and exclude it from our understanding? Why ignore the scriptures and the scientific implications hidden within them? As we've seen throughout this book, science and scriptures go hand-in-hand. God can use the BM-Fields within our bodies to reproduce the Life-Light of Christ within us. The "knowledge of God's Glory" cannot be reduced by religion and tradition in this age, for knowledge in the end times has increased (Daniel 12:4).

This knowledge of God's Glory is relational and revelatory, and it will help us to understand the science of DNA and all the elements of our biological system. With this knowledge, we will stay ahead of Lucifer.

Unfortunately, the religious leaders of this age are "always learning and never able to come to the knowledge of truth" (2 Timothy 3:7). The never-ending accumulation of doctrinal knowledge without relational experience of His glory makes for a dead religion with no face-to-face encounters with the King of glory.

The knowledge of glory is not esoteric knowledge. DNA has long been known to be a storage system for data, light, sound, and genetic language. It is being used by technology to store data. God will use it to store and duplicate His glory in man.

The knowledge of His glory is freely given, but many will reject the revelatory knowledge and end up following an empty religion that uses the same doctrinal knowledge minus the prayers and intimate encounters with God. Every individual must supplicate the Lord in earnest and develop a relationship with Him.

We are living in the most technologically advanced time in human history. It may be argued that the disciples of the first-century church, who witnessed the miracles of Christ firsthand and who were trained by

Christ to do the work of the kingdom, were the most privileged people in history. But I draw a case that today, with science and technology, we are potentially positioned to enter into an even greater era. We have greater access to the same revelatory knowledge of the glory because we can study the history of the church, and we can use that knowledge in tandem with the science and technology of the twenty-first century. We are truly blessed.

Advancing the Kingdom

The prelude to the end times and the manifestation of the sons of God is the understanding of God's glory. We have been given full access to the Father through prayer and worship. Through prayer, we can have relentless encounters with Him. Remember even as the cherubim and the living creatures are covered with eyes in Ezekiel 10:12, so are we covered with over 60 trillion eyes (cells) to behold His glory.

If we are honest, none of us are beholding the glory of Christ as we ought to, and we are significantly blocking the fullness of God from being revealed in our lives. The earth is groaning for the manifestation of the sons of God. Creation is waiting for the sons of God to release His Bioglory and Life-Light.

We can study the scriptures and Google almost anything on the internet. The science of Bioglory is not mysterious to our generation. Our failure to engage in spiritual warfare against Satan despite having this knowledge of Bioglory is the greater mystery.

Jesus lives to intercede for His church. He is always praying for us: "Father, I desire that they also whom You gave Me may be with Me where I am, that they may behold My glory which You have given Me; for You loved Me before the foundation of the world" (John 17: 24).

Bioglory Explosions

The church is about to experience a Bioglory explosion on the earth. The Father will do for us what He has done for Christ in the flesh! We will be anointed in the same way, through His sacrifice. We are being offered the same bliss. We are joined in oneness with Him—just as His life reveals you, your life reveals Him (Colossians 3:4).

We are in a co-morphic resonance with Christ. God has made us alive together through Christ. How can any human effort improve on this wholeness? The term *co-morphic resonance* (crucified and alive) with Christ should define our prayer lives. When you practice the prayer strategies in this book, Christ will be in you and you in Him (Galatians 2:19-20). As we have discussed in various places throughout this book, the work of Christ's earthly Bioglory ministry is finished and done. Christ in the flesh has completed our redemption, and we are able to sit with Christ "far above" all principalities and powers of darkness. Now the Holy Spirit is transforming us! Prayer-encoding Bioglory is about receiving all that God has for you, the newness of life in Christ—metamorphosis, immortality, everlasting peace, and joy.

In prayer, we can speak the Word of God into every strong nuclear force. The Word exits your mouth as a sword—quick, sharp, and powerful—to dismantle elements at a subatomic level. The Word has the power to destroy and realign every element (quark, neutron, or proton) that binds atoms together. The net result of all of this power is that we can use the prayer strategies in this book to accomplish His will.

Ezekiel 37:1-14 is a vision of hope. It contains a wonderful message that God's Bioglory can bring Life-Light to any situation and to any soul, no matter how utterly hopeless things might seem. Do you remember the story in which the dry, dead rod of Aaron came to life? After being placed in the Ark of God overnight, it budded, flowered, and produced almonds (Numbers 17:8). If God can make an old stick fruitful in His presence, He can do the same for us, both on a physical and spiritual level.

As a church, God can revive us and turn us into an army that will be part of His end-time remnant. Ezekiel 37:12-14 tells us that God's people will be raised up and brought into the promised land. A resurrection of the righteous will occur someday soon, and we will ultimately live in that new era in the New Jerusalem. But even before that, the Lord wants to raise up an army of spirit-filled people who will expand His kingdom.

Some of us do a pretty good job of covering up the "dry rods" in our lives. Maybe our marriages are dry and barren. Or maybe our family or work relationships have lost some vitality. Some of us have bank accounts that are dry and empty. Others have health problems. Whatever the cause of our desolation, the message in this study is that God can send you new life. God can breathe vitality through His Spirit and through His Word into our lives.

CLOSING THOUGHTS

Dear Reader,

I am glad to see that you have journeyed through this book with me. I hope it has been a blessing to you!

I firmly believe that what is coming has no historic precedent. This era won't be a mere repeat of past revivals or outpourings (as much as we love and honor the past). No mind-box can contain the limitless Bioglory of Christ that is within us. The Apostle Paul understood this truth; he said:

> The God of our Lord Jesus Christ, the Father of glory, may give to you the spirit of wisdom and revelation in the knowledge of Him, the eyes of your understanding being enlightened; that you may know what is the hope of His calling, what are the riches of the glory of His inheritance in the saints, and what *is* the exceeding greatness of His power toward us who believe, according to the working of His mighty power.
> Ephesians 1:17-19

I know we are growing in the knowledge of the glory of God. God is leading us into a molecular transformation from which we will never look back! The final generations on this earth are going to part take in the greatest glorious experience the world has ever known.

This is the greatest time to be alive. By praying with the language of glory, we will reveal the power and glory of heaven on the earth! And our power has been proven through science, technology, and biblical truths.

By accepting Jesus Christ as the one true Savior, by becoming born again in His Glory, by making time to pray, by bringing His Bioglory and

Life-Light into our spirits, by becoming transformed into the image of Christ right down to our DNA, by taking in His Word and sharing it far and wide, we can and will be changed.

www.ingramcontent.com/pod-product-compliance
Lightning Source LLC
Chambersburg PA
CBHW052041090426
42739CB00010B/1996